Kicking
GAS
& Taking
CHARGE!

DUANE A. LEFFEL

Clovercroft Publishing

To
Jenaia & Cayia
With sustainable love.

Kicking Gas & Taking Charge!

Published by Clovercroft Publishing, Franklin, Tennessee

Edited by Tammy Kling and Lee Titus Elliott

Cover and Interior Design by Suzanne Lawing

Printed in the United States of America

978-1-9425507-67-0

TO THE READER

I wrote this book at the encouragement of an Australian couple I met while traveling in New England in 2014; the wife told me I would always regret it if I didn't write one about my Guinness World Record cross-country adventure. I knew she was right.

In bringing the Ride the Future Tour story to life, I've tried to take you on the journey with me by using pictures from the Tour and describing the events from my perspective with candor, humor, and honesty. If I've done it well, you'll not only take an unforgettable ride across America; you'll also take a learning journey into sustainability as well. You'll come out of it wanting to do your part to keep the Earth as beautiful as it is today for your kids and future generations—just as I did.

To get the most out of this experience, I suggest you

1) Start by reading the character biographies in the Reference Materials section near the back of the book; you'll see just how different each of the Tour members were, coming from different countries, personalities, ages, backgrounds, and more.

2) Engage with our website kickinggasandtakingcharge.com along the way. Here, you will find additional photos, links, and information about the tour; a calendar of events related to the book; and interactive opportunities. I welcome your feedback on the book and the sustainability ideas presented in it.

Finally, I want to thank you for purchasing my book; I am humbled by it. And because I believe in "giving back," you should know that a portion of all profits from book sales will be donated to two causes close to my heart, namely the Make-A-Wish and Along Comes Hope organizations which provide support services to children with life-threatening diseases such as cancer.

Please enjoy *Kicking Gas & Taking Charge!*

ACKNOWLEDGMENTS

I want to thank John Arnesen for believing in our cause and for teaching me the basics about the LEAF and the electric-vehicle network. Without his persistence, help, and support, the trip and this book would never have happened.

I'd also like to thank AeroVironment and Nissan for lending me their products to make this journey.

To Susan Jones, I express my gratitude for having the foresight and conviction to make the Ride the Future Tour a reality. It truly was a once-in-a-lifetime experience that I will always remember.

Thank you to Larry Carpenter for educating me on the world of book publishing; your education and guidance was a huge help.

And finally, thanks to my parents for introducing me to the world of travel years ago and for giving me the drive to explore and solve problems, all of which contributed to this journey's success.

PROLOGUE

It seems every day we hear more stories about the fallouts of global warming—be it the polar ice caps melting, violent weather patterns or carbon dioxide levels. Most scientists believe the impact of fossil fuel use on the atmosphere is real. But even if you don't believe in global warming, the sheer impact of human growth and consumption suggests we need to find alternative ways to live and power our lives. Reducing internal combustion engines (ICEs) can improve the atmosphere, but there's much more to the word "sustainability" we should take an active role in.

This is the story about a group of individuals who decided to try to make a difference in the world by raising awareness of electric vehicles. Electrical vehicles are more popular in today's society, but in 2013, the idea of owning an all-electric car was still foreign to the general public. There were fears of exploding batteries and hazardous-material disasters; fears if the vehicles got into an accident; fears of being stranded along the highway if the all-electric cars ran out of power; even fears of having to take out a loan to pay the monthly electric bill to charge the car.

But what if a group showed that electric vehicles were safe, reliable, and economically feasible? That they provided an alternative to carbon-emitting ICE vehicles believed to be eroding our ozone-protected atmosphere? Could they change consumer mind-sets about electric vehicles?

This group of strangers took on the seemingly impossible task of crossing the United States using four different types of electric vehicles in the middle of a very hot summer, with no corporate funding and very limited planning. The goal: to increase public awareness and reliability of electric vehicles as alternatives to gasoline by setting four Guinness World Records, one for each type of vehicle. Few people have driven across the United States, let alone *ride* across the U.S.—and with electric vehicles?

The challenges were many, most of them unforeseen. The team encountered organization issues, bad weather, lost riders, power-source obstacles, injuries, and more, yet the team pressed on. Through planning and a lot of hard work, the team met the grueling forty-four day challenge with all the elements of a made-for-TV movie.

It's the story of a group *Kicking Gas & Taking Charge!*

CHAPTER 1

"What Have I Got Myself Into?"

It was a bit overcast in the early morning of July 4, 2013. Not a bad day, but it was early, and, given the recent weather pattern, rain was certainly in the forecast at some point. Yet we were all excited—ready to begin a journey full of questions. What challenges would we encounter? Would we get along together? Or would we even complete the trip? Uncertainty was what made it exciting. So we were ready. Or so we thought.

Most of us were expecting a crowd to see us off. But what I found at little Brittlebank Park in Charleston, South Carolina, where we assembled for our launch, were a few of Susan's friends and family, a couple of support folks for A2B Electric Bicycles, three riding enthusiasts who would join us for the day's ride, and a journalist. Maybe two journalists. But there were no TV camera crews. The only "crew" around was our own documentary crew of three. And where was the crowd lining the streets to cheer us on as we departed on this momentous first day?

In retrospect, maybe I shouldn't have had such high hopes. I mean, after all, so far the two preparation days in Charleston had been a lot of scrambling, questions, and confusion. Sure, everyone was friendly enough, but there was no real structure of responsibilities relative to who did what or when. No real, detailed plan had been laid out. Yet all had opinions on what they *wanted* to do!

The only thing that had been laid out was the route we would take. *That* was set because I had painstakingly marked out each mile from Charleston, South Carolina, to San Francisco, using Susan's southern

Dominique, Rachel, and Susan (i.e. the "scooteristas") line up in Brittlebank Park.

route. The route took highways from South Carolina, Georgia, Tennessee, and then Arkansas, before it went parallel to Interstate 40 and the old Route 66 from Oklahoma to Santa Monica. On paper, it was a trip that was 3,750 miles long (see Figure 1), and it was set to finish at Google headquarters in San Francisco, where we'd have a huge celebration. It was a trip that very few people have ever made by car and that no one had ever made with any all-electric vehicle, let alone four different types of electric vehicles. Yes, four types: an electric bicycle, an electric motorcycle, three electric scooters, and an electric car, the Nissan LEAF, which I drove. And all of us were hoping to set Guinness World Records for distance. This trip was a truly new frontier, and the route was important.

I didn't spend long hours and even all-night planning sessions with Susan deciding on the routes only because it was fun; I did so because I had to. I had to lay the routes out ahead of time because I needed to know where I could recharge the Nissan LEAF in order to set the GWR (Guinness World Record).

You see, there are roughly 121,500* gasoline-filling stations in the United States serving the millions of cars on the road today. However, there were only about 5,300** electric-charging stations spread across the United States for probably less than 30,000 electric vehicles. Here's a little better visual:

*U.S. Census Bureau 2012 **U.S. Dept. of Energy 2013

Figure 1

Our website: http://ridethefuturetour.com/

Gasoline: 1 filling station/ 31 square miles
Electric Chargers: 1 charging station/ 717 square miles!

While the average gasoline vehicle has a range of 375 miles per tank, my Nissan LEAF could only average about 95 miles per charge, depending on terrain and on how fast I drove it. And every recharge on 220 volts of (200v) power took three to four hours!

Knowing where every public and private electric-charging station was located was critical to my success and to our team's success. I had brought a gasoline-powered generator to charge the LEAF if I ran out of power along the way, but I didn't want to use it. That would show that the trip couldn't be done on electric power alone, and I didn't want that controversy. I wanted to make the trip entirely with electric power. That was the team's goal, and that was my goal.

Fortunately, there are websites and phone applications developed by the electric-vehicle industries that show where these power sources are located. The challenge for me was aligning our route with the available chargers and the LEAF's range. This may sound easy, but it wasn't. Though the major cities along our route from Charleston to Memphis had public charging stations, beyond Memphis was a vast sea of nothingness as far as public chargers go. It was like going to the moon. I didn't know *where* or *if* I'd be able to get 220 volts of power. So, I ask you:

Would you take off on a cross-country trip without knowing where the gasoline stations were?

Well, I was about to do just that. And, needless to say, it was a little intimidating. And scary.

As I sat in my LEAF waiting for my new friends to line up behind me, I reflected on all the confusion and sparse attendance I'd observed so far for the launch of this epic journey, and I remember asking myself:

"What the hell have I gotten myself into?"

CHAPTER 2

"That's a Cute Little Thing."

A couple years earlier, in 2011, I had gotten into hiking. My favorite spot was Radnor Lake State Park, a popular park just outside Nashville, where I could climb some semisteep ridges for a good workout and enjoy the scenery and wildlife. Fresh off a recent divorce, I used the hikes to clear my head and often rode my Goldwing motorcycle there to relax. Radnor was peaceful. I liked that.

After one of these two-hour hikes, I returned to my motorcycle to find a small scooter parked behind my 800 pound Goldwing. I thought the size difference was funny, and when the owner suddenly appeared, I made a snarky remark about it, something like: "That's a cute little thing."

That comment sparked a conversation. Susan Jones, the scooter owner, told me that it was an all-electric scooter and that she didn't own a car and rode the scooter *everywhere*. Since I knew the area, I knew she was riding on major roads and at some distances, so I was impressed. She further explained that she was riding the scooter because she wanted to show others that we don't need gasoline and reliance on the Middle East for oil. "Hmm.." I thought. "This spunky little blonde is not only cute; she also lives what she believes in." And after a lengthy conversation, we exchanged contact information before heading onto our separate ways.

Soon after, we began meeting for dinner or at events at her home, as we became friends. While we socialized a bit, the conversation always seemed to come back to her start-up electric-scooter business. Susan had procured the rights to be the sole distributor for Libert-E scooters from a company in China and created the company Xenon Motors to distribute them. Though she had no real sales outlet, she did have a few scooters and was picking my brain for how to best go about the Sales and Marketing aspect of the business, which was my career background.

Susan was a visionary. She wanted to sell her scooters without brick and mortar buildings (i.e., online) and without the middleman. Coming from the automotive industry, I recognized the challenges in selling her scooters, especially from a servicing-and-customer-satisfaction stand-point, and to implement her sales nationwide. Without brand recognition, such a plan would be extremely time-consuming. Susan began focusing on how she could get brand awareness and consumer acceptance of her electric scooters.

After some thought, I brought up the idea to Susan that maybe she could ride the scooter across the country to get attention; perhaps she could even try to set a Guinness World Record. I poked fun that she might get blown off the highway by some big trucks and end up in a ditch or something. To my surprise, she said she liked the idea and would be will-ing to make such a trip. A new seed had been planted, and Susan was off to start fertilizing and watering it.

Infatuated with the idea of eliminating all gasoline engines, Susan was very interested in the Nissan LEAF and decided to ride her scooter over sixteen miles from Nashville to Franklin, where Nissan's Sales and Marketing Headquarters were located. There, a Ride and Drive event was being held for the all-electric LEAF. Susan loved the vehicle and the event. Here, she met Jonathan Becker, a director who created marketing mate-rials and who had developed some of the service brochures and instruc-tion videos for the LEAF. They hit it off. And Susan's little seed got more fertilizer.

The LEAF now became a focus for Susan; she wanted it to join her in a cross-country trip because it would gain her more exposure for her scooters and for electric-vehicle awareness. And she knew somebody with inside Nissan connections—yep, me. I convinced her that if she wanted to use the LEAF in such an event, she needed to have a well-developed plan to convince Nissan to get behind it.

Susan began assembling a presentation package that outlined the concept, what would be involved, and how Nissan could benefit. I even arranged a meeting between her and Nissan Marketing Management, and she made her pitch. Susan did great, and she got some interest, but, with advertising budgets tight, the Advertising Agency didn't see how her pitch would fit in.

Needless to say, Susan was disappointed. But she continued to make some preliminary plans, and before long, she had prepared a more comprehensive pitch for Nissan. This one had more details on investment, as well as a focus on how Nissan would benefit from promoting a Guinness World Record. I got another meeting with Susan, this time with the vice president of Marketing. Once again, Nissan turned her down, stating that the message would be "inconsistent with their current marketing message to customers"; i.e., they didn't want to tell consumers they could cross the country because the lack of infrastructure and knowledge *could* result in customers becoming stranded without power—which would obviously be a black eye on the LEAF's image. Basically, the short-term risk was more important than the long-term benefits of promoting the LEAF as a viable alternative to gasoline.

Despite another crushing blow, Susan was determined to get a LEAF on her trip. But she also had other topics on her mind as she traveled back to Vietnam and Thailand in mid-2012 to finish a book she was writing.

For me, I was in the midst of the corporate rat race at Nissan. Citing cost reductions and communication benefits with the manufacturing sector, Nissan had moved the Sales and Marketing Headquarters from Los Angeles to Franklin, Tennessee, in 2006, and, in the process, the move left 70 percent of our staff in California. Making the decision to leave California was very painful for my family. After the move, most all of the former LA top management I had known had been swept aside and replaced with former Detroit Big-Three automotive managers or Nissan execs outside of the United States, leaving me one of the last of the LA executive management team left. The political and subsequent cultural changes were constant and frustrating. It's no coincidence that most directors and those ranked above them have left Nissan when they reach age fifty-five; they either fully retire, or they find a new challenge outside Nissan. Nissan is profitable and talented, but the company stretches their employees to the maximum to achieve financial targets.

I had a key date coming up in early 2013; I turned fifty-five in March and was contemplating a change. Work was no longer fun lately because of

- the loss of numerous coworkers I had known and worked with;

- new managers who thought they knew more than us veterans;

- changing internal HR policies. Our CEO's biggest fear was not the competition or the economy; it was "employees becoming complacent"; that pretty much said it all about the corporate HR climate.

Still, leaving a company after twenty-eighty years of dedicated service would not be easy. Nissan had provided a good career for me, and I still had loyalty to the company.

Just as I was contemplating my next move, I heard from Susan again. She hadn't forgotten about doing a cross-country trip on the scooter. In fact, she had a new vision: a small electrical "parade" that she named "The Ride the Future Tour." And through the Internet, she had recruited Ben Hopkins, a Briton with cross-continental bike-riding experience to ride an electric bicycle, and Ben Rich, a high school physics teacher from New Jersey who regularly attended motorcycle rallies on his Zero electric motorcycle. They counted themselves "in" and planned to join the Tour. But Susan still wanted a big draw—the Nissan LEAF. She asked me to join them and get her a LEAF.

The idea of driving the LEAF was intriguing, but I wasn't sure I was quite ready to leave Nissan, especially if I wanted to transition to another corporate job, because I didn't have another one lined up. In addition, I had no experience with the LEAF. Sure, I knew a little about it, but not much. The learning curve would be a challenge. I did, however, start making some inquiries to the Marketing Group about "what ifs" in the form of allowing the Tour to borrow or lease a LEAF for the trip. I got a little interest from John Arnesen, who was in charge of getting the AeroVironment charging stations into Nissan dealerships and the interface for the local Chapters of the Plug In America (electric-vehicle advocate) groups. I started learning from John, and suddenly it occurred to me that maybe I *could* make the trip successfully.

I began getting involved in Susan's videoconference calls, as well. The initial ones on the Google platform were a bit chaotic: she was trying to set up multiple video links between Bangkok, Nashville, New Jersey, and Hawaii simultaneously. However, we all eventually hooked up, and discussions about what would be necessary to put this journey together began. It was unusual in that the only connection that each person had was Susan, so all the participants had to introduce themselves and talk about their backgrounds. When my turn came, I was the guy trying to get the Tour a LEAF, but I hadn't committed, and I could sense the surprise from Ben

Rich (an EV enthusiast) when he found out I had no background in electric vehicles but was considering making an attempt to set a Guinness World Record. Ben was surprised, rightfully so.

During this time, I was having discussions with Susan about becoming her general manager for the scooter business. It was a ground floor opportunity—*really* ground floor. But I had done a fair amount of research on the possible sales and financials based on possible market share, and I was interested. I even went so far as to develop a brochure of the full product line that Xenon Motors would have as a "one-stop-shop" location for electric transportation. The Chinese company that Susan had an agreement with had a complete line of comparable gasoline-transport products that they had sold for years; they were just converting them to electric, as well. It had potential, and I could see myself running a company with these products. With the Ride the Future Tour launch date of July 4 fast approaching, I had to make a decision: was I in, or was I out?

The decision was seemingly easy, but it wasn't. I could stay frustrated at Nissan for a while longer, or I could take off on a possible once-in-a-lifetime trip across the country to try to set a Guinness World Record. Hmm. Ninety-five percent of the time I would have taken the "safe" answer and stuck with the company for a few years longer. But this time—*this* time—it seemed right to take a risk. "Why not?" I decided, "I'm going to do it!"

In mid-June, I let Nissan know I was set to retire and become head of a new electric-scooter company called Xenon Motors. A few days later, I was out the door. I still remember packing up all of my office belongings and leaving late on a Friday night. No farewell party. No cake. No celebration or even a "thank-you" for the twenty-eight years of my dedicated, hard work. One friend, Marty Gleason, walked out with me and my boxes. And that was it. Experience is not valued as it used to be, and, while sad, I knew I had made the right choice.

Before I left, I made the best of my time with John Arnesen. John gave me some much-needed insight on operating the LEAF, such as the types of EV charging (see Figure 2) and public-charging outlet vendors. He set me up with credit cards to pay for charging at the public chargers. He even lent me a gas can and an electric generator. It could be fired up to charge the LEAF if I ran out of power in some desolate location on the trip. Why was this valuable? Because, unlike gas vehicles, if the LEAF runs out of power, you don't move it. Rather you either recharge it on the spot somehow, or you flatbed truck it to a charging location, wherever 220 volts of power might be!

John, was a godsend. He was the one person who wanted to help out and see us take on the challenge.

John also connected me with AeroVironment (Nissan's charging equipment vendor), and we got two supplemental charge cords to use for my trip. I was referred to the NEMA (National Electric Manufacturers Association) plug chart to help me identify what types of electrical outlets I could tap into to charge the LEAF during my journey. To be honest, there were so many types of plugs and receptacles on the chart (see Figure 3) I found it very confusing to understand what plug type on the charge cords and corresponding receptacle I needed versus what I would *find* along my route. I didn't know if *any* outlets would fit the cords I had.

After studying the chart for a while, I realized that one of the cords would likely fit the outlets in Recreational Vehicle parks. So one day I drove twenty-five miles to an RV park in north Nashville and tried the

Figure 2: Types of Electric Vehicle Charging

The Nissan LEAF Owner's Manual gives a good, quick summary of the three types of charging the LEAF and other electric vehicles can use:

Types of Charge and How to Charge the LEAF LI-ON Battery

	Charge port / Charge connector / Power	Charging	Estimated charging time
NORMAL CHARGE See "CH. How to normal charge" in the "CH. Charging" section.	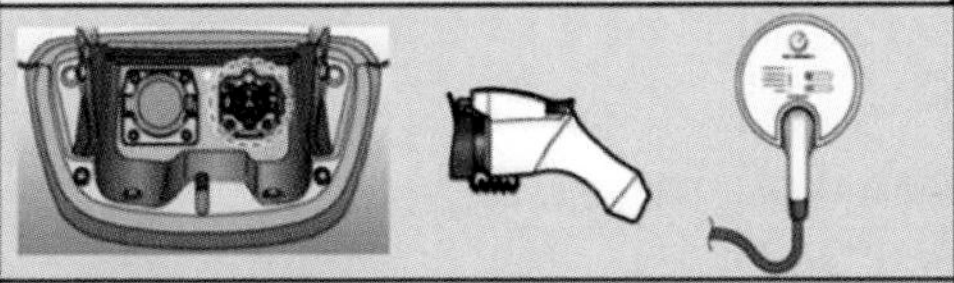	Use the charging device (AC 220-240 volt) that is installed in your home	Approximately 4-7 hours (depending on the charger) to charge the Li-ion battery from discharged (Low battery charge warning light illuminated) to charged.
TRICKLE CHARGE See "CH. How to trickle charge" in the "CH. Charging" section.	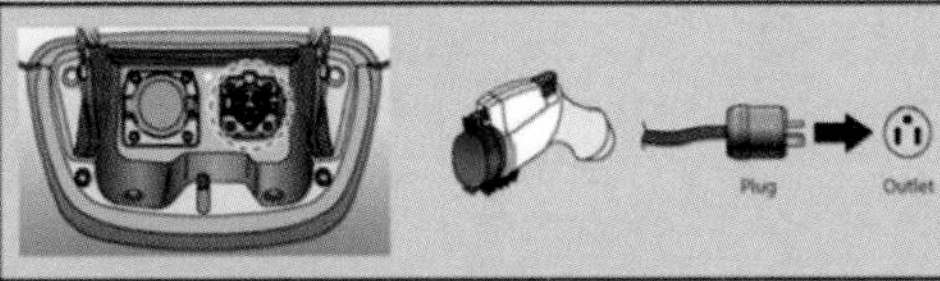 Plug / Outlet	Use the EVSE (Electric Vehicle Supply Equipment) or trickle charge cable for "opportunity" charging at a destination such as a friends house. Use only a 110-120 volt, 15 amp, dedicated outlet for charging.	Approximately 21 hours to charge the Li-ion battery from discharged (Low battery charge warning light illuminated) to charged.
QUICK CHARGE (If so equipped) See "CH. How to quick charge" in the "CH. Charging" section.	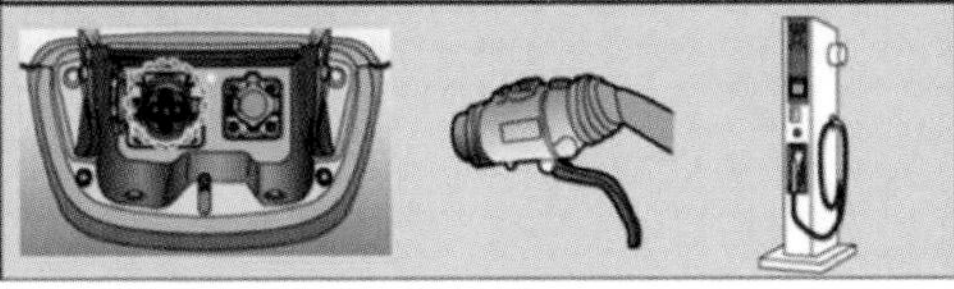	Public charging stations	Approximately 30 minutes to charge the Li-ion battery from discharged (Low battery charge warning light illuminated) to 80% charged.

The "Normal Charge" using a 220v power source is typically referred to as a Level 2 charge with the "Trickle Charge" on 110v being a Level 1 charge. The vast majority of public chargers in use today are Level 2 chargers, though you can find the "Quick Charge" CHAdeMO chargers using 440v at many Nissan dealerships

cord and plug in a 220v receptacle; it worked! I learned from a camper there that large RVs use 220v for running their big air conditioners and other electric appliances. This was a revelation! I now knew how I could get beyond Memphis for 220v power—through RV parks.

So just as I did for the public charging stations, I began locating all the RV parks along our route and overlapping them onto my Google Maps so I knew where I could recharge, particularly in Arkansas and beyond. This filled in a lot of holes between the Mississippi River and Los Angeles. Most, but not all. The rest I'd have to figure out along the way.

The LEAF itself was another issue. You would think that getting a car from a car company would be easy, right? Wrong. Getting a suitable car became a last-minute scramble with Nissan that involved three or four different LEAF vehicles. The first LEAF was a 2012; that was okay until I learned that the 2013 model had a significant improvement in recharging time and in driving range on a charge. I decided I had to have a 2013 LEAF to improve my chances of success, so I backed out of a vehicle swap agreement with a coworker. The problem was there were very few 2013 LEAFs available, and all of them were leased by employees!

Figure 3: NEMA Plug and Receptacle Chart

GENERAL - PURPOSE NONLOCKING PLUGS AND RECEPTACLES											
		15 AMPERE		20 AMPERE		30 AMPERE		50 AMPERE		60 AMPERE	
		RECEPTACLE	PLUG	RECEPTACLE	PLUG	RECEPTACLE	PLUG	RECEPTACLE	PLUG	RECEPTACLE	PLUG
125V	1	1-15R	1-15P								
250V	2		2-15P	2-20R	2-20P	2-30R	2-30P				
125V	5	5-15R	5-15P	5-20R	5-20P	5-30R	5-30P	5-50R	5-50P		
250V	6	6-15R	6-15P	6-20R	6-20P	6-30R	6-30P	6-50R	6-50P		
277V, A.C	7	7-15R	7-15P	7-20R	7-20P	7-30R	7-30P	7-50R	7-50P		
125/250V	10			10-20R	10-20P	10-30R	10-30P	10-50R	10-50P		
3Ø 250V	11	11-15R	11-15P	11-20R	11-20P	11-30R	11-30P	11-50R	11-50P		
125/250V	14	14-15R	14-15P	14-20R	14-20P	14-30R	14-30P	14-50R	14-50P	14-60R	14-60P
3Ø 250V	15	15-15R	15-15P	15-20R	15-20P	15-30R	15-30P	15-50R	15-50P	15-60R	15-60P
3Ø Y 125/208V	18	18-15R	18-15P	18-20R	18-20P	18-30R	18-30P	18-50R	18-50P	18-60R	18-60P

I arranged to swap vehicles with another employee who had just gotten a 2013 model—only to learn after we had traded keys that there was a problem with the insurance. I had company insurance, and the other employee had his own insurance coverage; the deal was nixed! Back to square one again. I was beginning to panic. The start of our journey was only a week away when John and the marketing group came through. They had a suitable white one that I could "drop off" on the West Coast for PR use there. It would work. Finally, I was "in"!

The next step was to get the vehicle certified by the local dealer as required by Guinness. Guinness has very stringent rules to follow when anyone attempts to set any world record, and I wanted to comply to the letter. The dealer did a number of inspections, checked the vehicle out as having no modifications, and validated the mileage on the car. The starting mileage was validated at 3,195 miles. I was good to go!

Susan, now in the United States, had gotten a custom wrap designed and had made arrangements for it to be applied to my all-white Nissan LEAF. I took the LEAF over, and within a couple days, voila! It was transformed!

The LEAF now had

- A sign announcing our attempt to set a Guinness World Record.

- A Ride the Future Tour emblem.

- A graphic map which showed our planned journey and cities across the country.

And it was cool!

My LEAF was now ready to be the lead vehicle for the Ride the Future Tour!

On Saturday, June 29, I met Susan in Nashville. We picked up the U-Haul supply truck we would need, as well as a trailer to tow the LEAF to Charleston, and we were on our way. Towing the LEAF by a trailer there

Starting mileage validated at 3,195 miles; note power-level bars and miles to go (right).

17

Nissan LEAF being checked for any modifications.

NISSAN of Cool Springs
Tennessee Automotive Group. Selection. Savings and Your Satisfaction

06/26/13

To Whom It May Concern:

This letter is to verify Nissan LEAF VIN # 1N4AZ0CP3DC403807 has not been altered and in the same state that is was when it left the production line.

Please feel free to contact me if you have any question.

Regards,

Mark A Hepler
Director of Fixed Operations
Nissan of Cool Springs

The local Nissan Dealer validated that the LEAF had original equipment. Good to go!

"Green Lightning" was born with the LEAF's custom wrap.

seemed logical to me because we were tight on getting to Charleston on time for the planned meetup on July 1. To drive the LEAF would have taken at least an extra day, maybe two days, because of the range limitations and the charge times. However, when we got to Charleston, I was informed that maybe my plan hadn't been a good idea. Someone had apparently snapped a picture of the LEAF on the trailer and had posted it to an EV website to claim that the LEAF wasn't self-sufficient—i.e., that electric vehicles weren't capable yet. It was almost as if I had screwed up. I remember thinking, "Really?" Was it further proof of the battleground between naysayers and electric advocates? Wow! This was nothing more than a distortion of the truth via social media. Now was the time to prove the naysayers wrong in black and white. "Game on!"

Handout card/sticker Susan had printed for us to build momentum during our trip.

Our Ride the Future Tour supply truck towing the LEAF en route to Charleston.

CHAPTER 3

Doing "the Charleston"

"The Charleston" is a dance named after the harbor city of Charleston, South Carolina, which became popular in the mid-1920s and the 1930s with the Broadway hit show Runnin' Wild. It is characterized by a fast-kicking step with both feet forward and then back, a sideways running-in-step while leaning forward, and a bow-legged quick passing of both hands back and forth as if you're trying to protect an exposed midsection of the body. Our few preparation days in Charleston were a lot like "the Charleston"—fast-stepping, running in place, and lots of trying to protect exposed areas!

Susan handpicked Charleston for the start of our journey because it was her hometown. With family and friends in the area, it was a good place for her to gather support and pull a string or two for press coverage. She set us up at The Mills House, a luxury hotel and historical landmark in downtown Charleston. It was here that, one by one, I met my compadres for the Tour.

Joining Susan and me would be

- Ben Rich, a high school physics teacher and electric-vehicle whiz and advocate;
- Ben Hopkins, a brawny Brit who had ridden bicycles across much of Australia, New Zealand, Europe, and China;
- Dominique, Susan's daughter and a pretty, young, aspiring singer-songwriter;

- Rachel, an attractive Aussie who loved children and studying cultures;
- Stuart, a lanky, active legislative lobbyist for carbon emissions and global warming;
- Sean, Stuart's son, a good-looking college student getting a biology degree and our support truck driver.

There were also three members of a documentary crew that Susan had reached agreement with late in June:

- Jonathan, the director who met Susan at the Nissan LEAF ride 'n drive event;
- Evan, the camera man who was young and ready to go;
- George, a veteran sound man who was pretty quiet and laid back.

The documentary crew were going to film everything we did for the entire trip; after the trip, Jonathan and/or Susan would market the film and submit it to different film festivals in hope of someone taking interest in our story and cause—perhaps someone like Tom Hanks who had recently done a film with Julia Roberts called *Larry Crowne,* which was about a scooter rider and who was supposedly an electric-vehicle advocate. Hey, anything's possible, right?

This was our team. We spent much of the first day, July 1, just talking and getting to know a little bit about each other. Later, we all went to dinner at a seafood restaurant called Hyman's, the owner of which was a friend of Susan's. It was great food. We had a fun time with the owner Eli and, in general, a great bonding experience for everyone before we embarked on this journey.

Tuesday was focused on logistics and planning. There were so many topics to cover in our meeting that it seemed to last forever. Things like

- **Where were we going?**
  Susan's plan was for us to drive a southern route to LA and then drive north to San Francisco and end our Tour at Google headquarters. Susan was enamored with Google because the company was very advanced in ecological programs and was rumored to be working on its own electric car.

Outside The Mills House: Kyle, Ben R., Sean, Rachel, Susan, Dominique, Ben H., I, and Stuart.

- **What roads do we take, and who would lead our little parade?**
It was decided that since I had laid out the routes, had a navigation system, and was in a four-wheel vehicle, I should probably lead the group. I had laid out every mile of our trip by city and by state with Google Maps and Recargo (for public chargers), so this decision seemed appropriate. (See Figures 4 and 5.)

- **What were the activities planned?**
Susan had developed a summary of the cities we would be stopping in along the route. They included some major cities like Columbia, Atlanta, Chattanooga, Nashville, Memphis, Little Rock, Albuquerque, Los Angeles, and San Francisco. It also included a lot of small cities and towns. But, in each of these, she had attempted to set up local events that would involve the public and local officials, electric-vehicle enthusiasts, the press, and more. It would mean, however, that, after our drive each day, we would be meeting the press and attending events. It looked like fun on paper because everyone would be involved. We'd get great exposure for our message, and we'd be able to interact with electric-vehicle enthusiasts. The activities she encouraged locals to set up were

> ➤ Electric-vehicle display and test-drive areas.

> ➤ Singing contests with prizes.

> ➤ Children's activities to learn about how to be "green" and why.

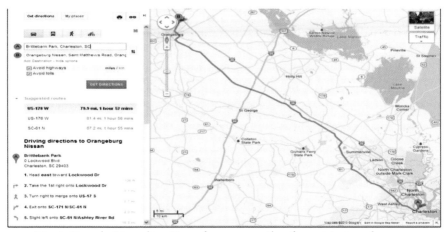

Fig.4: Google Map directions for every mile of our trip in advance.

Fig. 5: I also had copies of every public charger available along our route.

➢ Speaking opportunities for Stuart to talk about global warming.

➢ Opportunities for locals to ride along with the tour.

Susan had also created a website, ridethefuturetour.com, so along the way, more people would see our press, read our bios and mission statement, and get behind us by making plans for cities in the latter part of our trip. But, through all our planning, Susan hadn't had time to put all of the city plans together yet.

- ***Who would ride in or on what?***

The four Guinness Record setters were fixed: Susan, Ben H., Ben R., and I. Rachel and Dominique were going to ride scooters for most of the trip. That, however, left the three documentary crew, as well as Stuart and Sean, to ride in the supply truck. *Problem!* I soon realized I would have passengers for my trip, which I wasn't thrilled with because extra weight meant more power usage. Ugh.

- ***Where would we put all of our "stuff"?***

"Stuff" included a lot:

- eleven suitcases and travel bags,

- an extra scooter and extra bikes,

- spare tires and parts,

- tools,

- extra batteries,

- battery chargers,

- my gas generator,

- eleven sleeping bags and tents, as well as camping equipment,

- camera equipment,

- musical instruments,

- banner-making materials and crayons,

- food and snacks,

- water and drinks and, of course, beer!

All of this "stuff" had to fit in the supply truck, unless of course, the LEAF could haul supplies too? However, it sounded like I was already going to be hauling some of the extra passengers, and further added weight was going to be problematic. I pushed back, stating that I felt the added weight was going to compromise my attempt to achieve the GWR, given I had a severe range limit and would need all of my power just to get from charge station to charge station. So it was decided that all of the "stuff" had to go into the supply truck.

Stuart and Susan watch as Sean tries to locate something in the supply truck.

- **What were the sleeping arrangements going to be?**
  Susan had us paired up at each hotel we stayed at. The pairing was her choice and subject to change. No one had an issue with that. In a couple of locations, we would be staying in local houses, camping in state parks, and one night staying at a commune (because the commune residents had offered it to Susan for free).

- **Where would we eat?**
  Susan had packed some snacks for us, but she would also give each of us a weekly food stipend of $200. It wasn't a lot, but you decided how you wanted to use the money for breakfast, lunch, and/or dinner.

- **How would we charge and exchange batteries?**
  In general, batteries would have to be charged at night. These included the small batteries for the bicycle and the scooters, as well as the large ones for the LEAF and the motorcycle. The LEAF and the motorcycle would need to recharge together because they used the same public 220v charging outlets; because the mileage range was similar for both; and because of their speed.

- **How would we communicate?**
  Everyone but Ben Hopkins had cell phones, so we exchanged numbers. Ben, being from England/Thailand, was going to rely on A2B Bikes to get him a prepaid phone to use. We also purchased a walkie-talkie set, just in case.

After all this discussion, we thought we were ready. Wednesday was our final coordination and packing day as well as an opportunity for Susan to show us a few local sites around Charleston. I learned that the documentary crew had apparently volunteered to go to Home Depot to buy and assemble some shelving for the supply truck; this would hopefully provide some organization for the chaotic state it was currently in. Everyone else was scrambling on various last-minute tasks.

For me, I wanted to make sure I had everything I needed for the Guinness Record attempt. I spoke to the others who would be attempting a GWR, but there wasn't a cohesive understanding of what was necessary.

Ben H. was relying on A2B Bikes to do whatever was necessary for Guinness. Even though Ben would own the record, A2B Bikes wanted to use the promotion opportunity to tout the company's electric bike's durability.

Ben R. was going to rely on his speedometer for recorded miles, but he was worried that another group was circling the interior of the United States with electric motorcycles, which would negate his attempt at a record.

Susan hadn't prepared much other than the initial GWR application for all of us. She had a few other things on her plate though—like *everything*!

I had carefully read the Guinness Guideline Pack. Guinness has very strict guidelines for setting world records to ensure accuracy and fairness. In their words:

> "These guidelines are specific to your attempt and **must** be followed. Should any guideline be contravened, your attempt will be disqualified, without any right of appeal."

I wanted to follow the guidelines to the T. But even though they were strict, the guidelines weren't really clear on what's required, as I saw it. For my all-electric (nonsolar) vehicle attempt, the primary guidelines were

* A commercially available solely electrically powered car	✓
* The car may not be altered in any way from production line	✓
* A professional mechanic must check the vehicle before the journey to confirm the unaltered state and submit on company letterhead	✓
* Accurate professional equipment (GPS) and associated printout must be used and kml files submitted as evidence (for mileage)	??
* Details calculated distances must be given before the attempt starts so GWR can confirm the measurements are correct	✓
* Participant cannot remain stationary (same location) > 14 days	✓
* Legs of the journey must be continuous (i.e. each starting point is at the end of the previous leg)	✓
* Witness Book showing Date & Time, Location, Name, Signature for independent witnesses of the event to sign	✓
* Log Book providing an adequate description of the event and full details of participants daily/overall performance	✓

I was comfortable that I was set for everything except for the guideline about the GPS on the distance tracking. I would have GPS in the car and record all of the speedometer readings at the end of the day; I thought that should be accurate enough given the car inspection I'd had. But the words "professional equipment" and "printout" were haunting me, and it was too late to call England to verify. I went to Home Depot with Ben Hopkins and found a GPS system that would track longitude and latitude locations along our trip. And it could be printed out. It was a last-minute decision, and I didn't have much time to figure it out, but I bought it. This turned out to be a wise decision for my official GWR application and mileage validation.

Late Wednesday afternoon and evening, Susan took the group over to Sullivan Island, a historical entry point for the African slave market and a military fort used for defense of the city during the Civil War. We then went to a nearby beach, where the documentary crew wanted to film us for a trailer that we would post on the website. It was early dusk, and we all lined up side by side looking out at the ocean. This was our send-off: a look to the Atlantic Ocean before we headed to the Pacific. We stood there excited, chatting, and reflective. We were beginning to feel like celebrities. We were the focus of attention. As a few beach-goers watched, the cameras rolled. And we would continue to be the focus for forty-four days.

Lined up on the beach for our initial documentary coverage.
This scene was later used for our documentary teaser:
http://ridethefuturetour.com/documentary

CHAPTER 4

Day 1: Charleston to Columbia

"There's too much Fatz in our Daily Plan!"

So in the early morning hours of July 4, 2013, we headed off from our hotel to the launch point. Brittlebank Park is a small park on the Ashley River flanked by the downtown home of the Charleston Riverdogs minor league baseball team and the Coastal Carolina Yacht Sales in Bristol Marina. There isn't much in the park itself except the cul-de-sac parking lot, a grassy area leading to the river, and a few kids' play areas. But we had to start somewhere, and this was the spot Susan had chosen to begin our journey.

I didn't know it at the time, but most of our trip would be like this—little fanfare and turnout. However, those we did talk to were almost always interested in our endeavor—especially the electric-vehicle community. But here in Brittlebank Park on July 4, it was mostly Susan's friends and relatives that were seeing us off. Well, it was a start.

We spent a couple hours here with the reporter, Susan's family and friends, and a few curious people who wandered by. Susan, as well as Kyle from A2B Bikes, met with three people who came to ride with us for the day using extra electric bicycles A2B Bikes had brought. It was fun to have people want to join us because they were always enthusiastic about what we were trying to accomplish. I would ask these folks to sign my logbook each time we reached our destination.

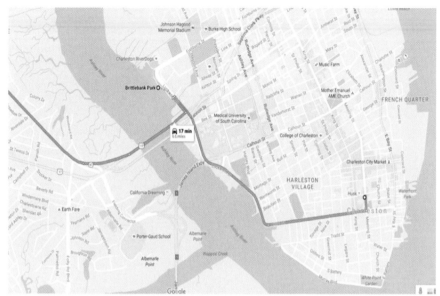

*Map of downtown Charleston showing our path
to Brittlebank Park—our kickoff point.*

Our target for the day was to drive to Columbia, which was about 120 miles from Charleston (see Figure 6). In a car, this would take about an hour and fifty minutes. For us, it would take all day. For me, since the LEAF could only do about ninety miles on a full charge, it meant a recharge in the small town of Orangeburg, where there was a Nissan dealership with a charging station.

Dominique makes a final music check.

Ben R. speaks with an enthusiast.

Date		City	St	Distance To Next Stop (mi)	Re-Charge Location	Contact Name	Phone Number
Thursday, July 04, 2013	Start	Charleston	SC	79.9	Charleston Visitor Center		1-800-774-00(
	Pit Stop	Orangeburg	SC	41.9	Orangeburg Nissan	Carl Payne	803-534-525
Total				121.8			
Friday, July 05, 2013	Start	Columbia	SC	69.4	Electric Cooperatives-SC		803-796-606(
	Pit Stop	Union	SC	53.6	Town Hall	Debbie @city hall	864-429-170
Total				123			
						Kevin Styne	864-467-448
Saturday, July 06, 2013	Start	Greenville	SC	30.3	Greenville Zoo		864-467-430
	Pit Stop	Anderson	SC	69.5	Piedmont Nissan	Audrea	864-328-118
Total				99.8			
							SKYPE
Sunday, July 07, 2013	Start	Athens	Ga	69.7	Nissan of Athens	Jennifer	706-549-660
	Pit Stop	N/A					
Total				69.7			

Fig. 6: Portion of a daily schedule tracking distance and LEAF recharging locations.

Scooter riders prepare to ride, and later the scooteristas fall into line.

Finally, we were off, with me in the lead; Ben R. on the motorcycle; the girls—Susan, Dominique, and Rachel—on the scooters; Ben H. on the bike; and Sean in the back with the supply truck protecting our rear from faster traffic. Our first challenge was navigating a bridge across the Ashley River, which we did. But as we headed north on the highway, it became apparent that traveling together wasn't going to be easy. Vehicle speeds, traffic lights, impatient drivers, weather—all became factors.

While the LEAF, the Zero motorcycle, and the supply truck could do normal highway speeds and stay with traffic, the scooters maxed out at 35 mph, and Ben H. on the bike could only average 15–25 mph. We had discussed these issues in our meeting beforehand and decided that we would stay together initially. We traveled basically single file on a two-lane highway out of Charleston, but the road was not wide enough for traffic to pass us since oncoming traffic was headed our way. So we stood our ground and stayed in line. But as some traffic did get around us, the drivers were sometimes honking their horns at us, obviously frustrated in our delaying

31

them and really not caring about any Guinness World Record.

We kept traveling. But we suddenly found ourselves being followed by some of South Carolina's finest, with flashing blue lights. We had barely gotten out of Charleston, and already we had been pulled over by the police. We were expecting a ticket, but, fortunately, the police officer only gave us a warning. Apparently, some drivers had been ticked off enough that they called the police about our slow pace. We were told we had to pull off the road more often to let traffic by. A bit relieved, we continued. At the same time, everyone must have been thinking about being one day into the mission and already pulled over by the police. Ugh.

Because Ben R.'s Zero electric motorcycle also had to recharge, it was agreed that he and I would go ahead of everyone else, recharge, and then catch back up with the rest. That was the plan, but throughout this trip, we found plans would often change.

Ben R. and I found the dealership in Orangeburg easy enough, around 1:00 p.m. Fortunately, the charging station was in the front of the dealership; its location had been a concern since it was a holiday and no one was around. The bigger issue was there was only one charger; only one of us could charge at a time. This was a setback because potentially it could take us three to four hours to both recharge enough to make the remaining forty miles to Columbia. Double ugh.

We started the charge on my car and walked up the street to a local café called Fatz to get a bite to eat. Meanwhile, we learned the girls and Ben H. had gotten drenched in some torrential downpours , so they were behind schedule. They decided to meet us at the restaurant and arrived in the early afternoon.

Apparently, no one had remembered to charge the scooter batteries the night before, so the scooter riders also needed to charge batteries while they ate. Soon all the outlets in the restaurant had battery chargers hooked up to them. We ate, drank, and had fun at Fatz, but with these two recharging delays, we were now way off schedule. There was a July 4 street-fair celebration in Columbia, where we were to be featured attendees. And we'd be late.

After spending most of the afternoon at Fatz, we finally got on the road again about 5:00 p.m., arriving in Columbia around 7:30 p.m. There wasn't a lot of July 4 celebration left by the time we arrived because of the earlier rain and our lateness. Undaunted, we parked our vehicles and talked to

Left: Ben R.'s Zero motorcycle on charge.
Note that only one vehicle could be charged at a time.
Right: Scooter batteries charging at Fatz Restaurant.

the few people still there.

Rachel tried to get some activities with the kids going while Ben R., Susan, and I talked to local organizers and passersby. Ben H. and the other bike riders whom we'd left behind arrived around dusk.

Those still in the area and curious came by to chat. But overall, with the crowd limited and most activities wrapping up, the event was disappointing.

We stayed until dark and then headed for the hotel. However, I had to drop off the LEAF at a charging station along the way, so I ended up riding one of the extra electric bikes a few miles to the hotel. This sounded easy, but it was dark now, and the scooter lights were dim; the bicycles had no lights; and we were traveling on a very busy street. Then, we started up a very steep hill.

This quickly became very scary for all of us because traffic was going much faster than we were. To make matters worse, I found myself having to pedal far more than I should have—for some reason. The pedaling was so strenuous I couldn't keep up, so I pulled off on a side street. We regrouped in the dark and waited for the supply truck to double back so we could try to fix the bike. There were no streetlights nearby, and it

Ben H. poses for pictures with Kyle from A2B Bikes and our two guest riders who completed the day's trip.

Rachel gets the kids involved in "green" activities.

A late July 4 celebration in Columbia.

was almost impossible for Kyle, from A2B Bikes, to see and fix the bike. So we loaded it into the truck, and I got in the back of the van in pitch black as we drove to the hotel.

The next day, I found out that I hadn't been shifting properly and had broken the shifting mechanism on the handle. I felt bad about breaking the shifter, but it was an accident. I was just glad that we all got out of that potentially dangerous situation—i.e., being in the dark with no lights and traveling much slower than traffic around us. Cars were coming up on the bikes and scooters far too quickly, and it would have been easy for someone to plow into one of us. This little episode made us all unsettled. We met briefly back at the hotel to discuss the day's events and make adjustments, but it was approaching midnight, and everyone was exhausted, and it was only the end of Day 1.

CHAPTER 5

Day 2: Columbia to Greenville

"Never has an event been more secure."

Because we'd gotten in late the night before, there were some adjustments being made in the early morning, so Day 2 got off to a slower-than-expected start. The drive to Greenville would be about 105 miles, and we had a consumer "green" event at Furman University that evening. So not getting started until almost 10:00 a.m. was probably not a good sign.

Once again, the LEAF and Ben's motorcycle would need to recharge along the way, this time in the small town of Union, which was some thirty miles out of our way but the only place with a Level 2 charger in the area. So our trip was 130 miles for the day, and we likely wouldn't see much of the scooters and bike.

Stuart rode with me today, and as soon as we retrieved the LEAF from the overnight charging station, we took off for Union. We arrived around 11:15 a.m., found the charger, and connected up. Soon we were met by one of the clerks from the building nearby who had been waiting for us. We'd called ahead of time to make sure the charger actually worked (I had learned from John Arnesen that sometimes they did not), so when we arrived, we were greeted by our new friend and a reporter. The reporter interviewed us about our adventure and took several pictures; they later appeared in the local paper with an article about us.

Since there were two working chargers here, our recharging time wouldn't take as long. We headed off to have a nice, long lunch, and upon our return, we were given some Union-labeled drinking cups and an umbrella each. It was a nice gesture, and we appreciated the hospitality. Between the hospitality and the larger-than-life mural painted on a nearby building, I won't soon forget Union, South Carolina.

We continued on our way and reached Furman University at about 4:30 p.m. Shortly after we arrived, however, there was another nasty thunderstorm, and we took refuge in one of the university buildings. In July 2013, the Southeast was deluged with rain. For example, the Greenville area had over seven inches of rain in the first six days of July that year when usually the same period averaged less than one inch! Rain was already starting to wear on us.

The scooters and bike were running late again and had gotten wet, so it would take them awhile before they arrived at Furman. We decided to check out the campus a bit and managed to find a couple of the professors involved in the university's sustainability program. They gave us a short tour and interesting information on their efforts to be fully independent of outside power and, in fact, have their own power generation on the grounds. Furman is the only liberal arts institution in the nation to offer a bachelor's degree in Sustainability Science and one of only thirty-eight schools to achieve Gold Status in the Sustainability Tracking Assessment Ratings System (STARS)! As an alumna, I'm sure Susan is very proud of this accomplishment.

Meanwhile, the time for the evening event where we would be talking

Town folk greet us, as Ben R. and Stuart
wave during our recharging in Union.

*"Watch out !" No it's not an actual train
about to hit an SUV; just great artwork!
www.cityofunion.net*

*Talking sustainability with Furman professors (and their families).
www.furman.edu*

to the public about our efforts and discuss subjects such as electric vehicles and sustainability was fast approaching. We saw several security guards walking the parking lot, ready to direct traffic and ensure public safety. We drove over to the amphitheater, where the event was to be held, to see the crowd and found, well, nothing. Not 1,000 people, not 100 people. In fact, not one person was waiting for us to talk to! "Well, this was embarrassing."

While the security guards held down the fort, we let Susan know that the night was obviously a bust, and we went on to the hotel. Whether it was an organization issue or a rain issue, I don't know, but I do know that "never has an event been more secure" than our Furman event that night. Oh, and by the way, even if you're Furman alumni, there are "no refunds" on events held.

CHAPTER 6

Day 3: Greenville, S.C. to Athens, Ga.

"From Terrapin to 'where's Ben been?'"

July 6 was scheduled to mark our third long drive in a row—at about 110 miles. One of the locals had suggested that we might want to go to the Greenville Farmer's Market downtown on this morning because we could talk to people about our electric vehicles as they walked by. So early Saturday morning, Benswing (two Bens was getting confusing, so we came up with Benswing for Ben Rich as a nickname for his swing dancing) and I drove down to the Farmer's Market and attempted to find a spot to set up shop. However, Security soon advised us that we couldn't display on the main street where all the shoppers were; we'd have to park on one of the side streets and hope people walked by.

After about an hour of minimal interaction, we decided to be on our way to Athens. Our first stop, however, would be Piedmont Nissan in Anderson, South Carolina, where we would recharge. Almost all Nissan dealers had Level 2 charging equipment, but its accessibility was usually a function of the priority each dealership assigned to electric vehicles. We were usually greeted with open arms to come into the store and relax. Some salespersons and other employees got excited about the LEAF and wanted to have pictures taken with us. A few ignored us. Most of the time, I didn't tell them I was a former Nissan executive just to see how they treated me. More important, I wanted to know that they saw me as a

Ben R. loads his camera from the LEAF with the Greenville Farmer's Market in the distance.

customer—somebody who could help their business with LEAF sales if I was successful.

Piedmont Nissan was very gracious (I think they liked Rachel's Aussie accent) and even shuttled us to a nearby favorite eating establishment for lunch. We took off again for Athens about 1:00 p.m. and arrived early at our event location for that evening, the Terrapin Brewery. But, along the way, we once again ran into rain, which meant another delay for the scooters and bike riders.

A brewery seemed like an odd venue for our project, but somehow Susan had gotten a connection to it. I'm not much of a beer drinker, but I do find it interesting that Terrapin bills itself as a "craft brewery," a specialty beer company, and that the brewery embraced Athens as its home because of its similar "love of music, commitment to the environment, and practice of living life to the fullest." Terrapin interacts with its customers regularly by booking events on the property adjacent to its brewery. In essence, everyone comes in, drinks some beer and eats with friends, interacts with the booked attraction, and goes home. And on this night, the Ride the Future Tour would supplement the planned music! Finally,

Recharging at Piedmont Nissan.

Terrapin Brewery's customers had a good time drinking, eating, and hearing about our adventures and cause.

we might get to talk to some people.

Benswing and I pulled our vehicles in for display and spoke to the folks meandering around in between Frisbee tosses and dog walks. The girls and BenHop (our nickname for Ben Hopkins) finally arrived to fanfare, and soon we had a microphone talking about our backgrounds and the Tour objective. The audience was courteous and supportive as we talked. And with the beer, everyone had a good time!

As the event was winding down, we began to plan our trek to an exchange house that Susan had booked us in to stay for the night. We soon realized though that we had a problem: Terrapin Brewery was nowhere close to the house nor close to downtown where the Nissan dealership was for me to recharge the LEAF. Fortunately, we had a special bike rider with us that day, Major Andrew Lane. Andrew and his wife graciously invited us to dinner at a local restaurant and offered to shuttle me from the dealership to our accommodations on the other side of town. We were all set once again, thanks to friends like the Lanes.

Andrew, "Captain Planet," as he calls himself, is all about sustainability, solar power, electric vehicles, and recycling. The guy lives and breathes all of it. He wants to make the Earth a better place to live. But as we started out on the trip from Greenville, Andrew with his bike and big canvas-container trailer was stopping along the roadside to pick up every aluminum can he spotted. I mean, he took this task very seriously, and I thought it was going to take us *all* day to go thirty miles!

But Major Andrew lives his life as he believes, and he was a *huge* supporter of our efforts. Courteous and respectful in all he does, he continues

View across top of the LEAF of Terrapin Brewery;
Rachel and I talk to attendees.
http://terrapinbeer.com/

Terrapin's commitment to sustainability is evident
throughout the brewery.

to communicate with us today as he carries on in his mission to be fully sustainable. The world needs more Andrew Lanes for our planet to survive.

"Thank you Captain Planet. I salute you!"

After a nice dinner, we split up. I headed to the dealership and the others toward our house for the evening. But just as I was getting to the dealership, another nasty thunderstorm hit. Fortunately, I got connected okay, but I had to wait for Andrew and his wife because they had the kids to deal with before me. So I sat at the dealership, waiting in the storm for my ride.

The riders weren't so lucky. They had gotten caught in the storm and had taken cover. But as soon as there was some letup, the scooters went on crossing flooded streets, and the riders were getting drenched, but, in the process, they left BenHop behind. I finally got to the house about 11:00 p.m., but I learned that Ben Hopkins wasn't there yet. We were all wondering, "Where's Ben?" And we started to worry about him because he hadn't contacted us. He finally arrived around 12:30 a.m. after fixing a flat tire and getting soaked to the bone. It was a very long day for all of us, but no one's was longer than BenHop's.

Major Andrew Lane

CHAPTER 7

Day 4: Athens to Atlanta

Aristotle: "Worms are the Intestines of the Earth."

(taken from the Worm Dude Website; www.thewormdude.com)

We got a break from the long drives on Day 4. Our destination was Atlanta, which was only about seventy miles from Athens. This meant no need to recharge in the middle of the day. While the girls and BenHop slept in and recovered from the previous night, Benswing, Rachel, Stuart, and I decided to go over to nearby Stone Mountain. Stone Mountain, if you don't know, has the largest relief (into mountain rock) sculpture in the world. This Confederate Memorial Carving depicts three Confederate heroes of the Civil War, President Jefferson Davis and Generals Robert E. Lee and Thomas J. "Stonewall" Jackson. The entire carved surface measures three acres, larger than a football field and Mount Rushmore. With the carvings, there is also a museum and beautiful scenery all over the mountain.

After a relaxing couple of hours at Stone Mountain, we headed on to Atlanta; there was a consumer event planned for 4:00 p.m. there. When we arrived we found the A2B Bike folks and a couple of people who owned other makes of electrical cars. And a little later, two Nissan LEAF owners showed up.

While not a large crowd, these were the "early adopters"—i.e., people who believed in electric vehicles as a primary transportation medium and

Rachel, Stuart, Benswing, and I on the tram to the top of Stone Mountain
www.stonemountainpark.com/activities/attractions.aspx

who bought electrics when they were first available. Our conversations were focused not on the vehicle potential but on performance, problems, concerns, and more. One topic that came up was battery life. Batteries were a big concern to owners because new ones for electric vehicles can be one-third to one-half of the price of a new car and initial estimates were that batteries would only last for seven to eight years. Manufacturers have seen longer-than-expected battery life, so we of course passed this information along.

Our event had been staged in the parking lot of the Southface Energy Institute. Southface is a leader in the Southeast in sustainability and promotes sustainable homes, workplaces, and communities through education, research, advocacy, and technical assistance. They are uniquely positioned to connect industry, government, and nonprofit sectors to develop sustainability solutions. Their Eco Office showcases more than 100 off-the-shelf technologies and design techniques that can improve customer sustainability, including residential opportunities.

We were given a guided tour of the roof, which includes dry-area plants, a water tower to collect and store rain water, and solar panels. And while the interior looked like a regular office building, it contains many of

Electric motorcycles were on proud display outside our Southface Energy consumer event.

LEAF and other electric car owners stopped to talk about the electric experience.

the technological advances to make the building totally independent for power, utilities, and water.

One of the most interesting operations was vermicomposting, or the composting of kitchen food waste by worms; it makes nutritious food for plants. And a recently conducted study included vermicomposting toilets, which use worms to dispose of human waste also. (I didn't ask what type of plumber you would call if you have a leak in that type of toilet system.) The tour really makes one think about what is possible for sustainability. And after seeing their huge bin of earthworms eating human food waste, Southface certainly proved to me that Aristotle was right: "Worms are the intestines of the Earth"!

Later that evening, I went looking for a place to recharge the LEAF, which took me to Lenox Square Mall in Buckhead. I was looking for the rechargers when the mall cops rolled up. They wanted to know if they could help me. I explained I was there to recharge the LEAF. I was informed that the mall was closed and that I needed to leave. I reiterated the importance and need for me to recharge, but it fell on deaf ears. Not very good customer relations, Lenox Robocops!

We drove to a nearby Walgreens, which was also shown to have chargers, and, sure enough, there was one there. We plugged in, and I asked the manager if I could leave the LEAF there overnight, and he agreed. (Much better customer relations!) We went back in the morning, as promised, and away we went.

I did learn that, in the state of Georgia, every Walgreens has installed a

Stuart checks out the plant life of the roof of the Southface building.
www.southface.org

recharger for electric vehicles and it's available 24/7. Public "shout out" to Walgreens in Georgia:

"I'm glad you are ahead of the game for electric power. Nice job!"

We need more companies to be proactive in alternative fuel sourcing. It's not only good for the environment; it can also be good for business!

CHAPTER 8

Day 5: Atlanta to Calhoun

"Baseball, Apple Pie, and a Blind Eye."

Monday, July 8, was scheduled to be a rather easy day. It was only about seventy miles to Calhoun, and we were to do a consumer-interaction event at a Little League Baseball Tournament there at 11:00 a.m. Benswing and I left around 8:30 a.m. and found our way to Calhoun; however, we didn't find a baseball tournament anywhere. We asked around and found out that the tournament was in Somerville, a nearby small town. On we went to Somerville, but still no baseball tournament. We did find the high school where it was to be played and learned that the games didn't begin until midafternoon. So back to Calhoun we went, Ben and I, to the Level 2 chargers in the center of town near the municipal court and began to repower. To kill time, we walked down Wall Street and found B&J's Diner for lunch.

Walking into B&J's was a little unsettling, actually. Inside were a number of loyal patrons who eyed us curiously; it seems B&J's is a favorite lunch spot for a number of the local police force. So either B&J's had good food, or B&J's had designed a smart security system for its little diner.

Before long, Barbara (of Barbara & Jimi's) had greeted us and took an interest in our trip. We had a great conversation and some good home cooking at B&J's. And ironically, it was here I found a replica of an early 1910 advertisement for an electric car—on the restroom wallpaper of

Recharging in Calhoun and enjoying the music in the Municipal courtyard.

all places! It's true: electric cars have been around a very long time. The LEAF, however, was the first to be mass produced and to offer many of the requirements that buyers want.

When we returned to Somerville, we hung a banner between two poles and set up our electric parade of vehicles on a patio where fans would have to walk right by us to get to the bleachers for the game. It was very hot and muggy, but we stood out in the sun and greeted folks going into and coming out of the tournament. Dominique sang songs, and we had our vehicles on display to talk electric.

It seemed like a nice idea, but we found that folks headed to a base-ball tournament to see their kids play don't really have a lot of interest in "salespeople" or distractions. It appeared that some of them were late and in a hurry; others simply turned a blind eye to us as they went by. The Tour group had to temper their enthusiasm after a while because few people stopped to talk. Eventually, we packed it in and headed to the motel.

I'm sure it was a great tournament for a Monday, but my kid wasn't playing, so I didn't stick around, either.

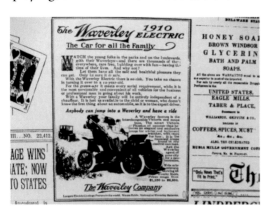

Where do you find the best reading materials? On restroom walls of course!

CHAPTER 9

Day 6: Calhoun, Ga. to Chattanooga, Tn.

"Fix Me Please, SUVs, and SOBs."

On Tuesday, the documentary crew rode with me for the day. It would be a short ride, a little less than seventy miles to Chattanooga, and the ride would take us through part of the Chickamauga and Chattanooga National Military Park, a Civil War battlefield.

Before we left Calhoun, however, we had to make a stop. One of the documentary crew had read about a place called Sam's Tree House in Roadside America, so we had to check it out. This unique house is owned by, was built by, and is now lived in by Sam Edwards—a well-known author and adventurer, as well as a former aide to President Jimmy Carter. The house is literally built around a tree that integrates

- a submarine prop from one of Elvis Presley's movies;
- an airplane used as a bedroom;
- a helicopter;
- a boat;
- a den, a library, and a kitchen.

An American flag is flown proudly at the "front" of the house for good measure. So the next time your spouse wants to "remodel" your house, suggest something new and different and offer this up!

As we continued on, the documentary crew had to video us as we

Rachel and Benswing take in Sam's Treehouse while Evan videos it.
www.samstreehouse.com/upatree.html

crossed into Tennessee; they did this whenever we entered a new state. It marked a sort of celebration of another milestone for us. Each of us pumped a fist as we drove by the camera crew, cheering.

We were now in some familiar territory for me because Nashville was only two hours away from Chattanooga. Since we had a little extra time before our late-afternoon event, I suggested that we go to the Lookout Mountain Incline, which is just outside the city. We all went along with the idea, and up we went!

The Lookout Mountain Incline Railway is a one-mile long narrow gauge rail line to the top of Lookout Mountain. A Historic Mechanical Engineering Landmark, the railway has a maximum grade of 72.7 percent and is recognized as one of the world's steepest passenger railways.

There's an observation platform at the top with a great view, so we paused to take it in. Interestingly, Lookout Mountain was the scene of a major battle during the Civil War, which was fought on the mountainside. There's a national park with cannon and battle memorabilia a short walk from the railway.

As we began descending back down on the railway, I became focused on my new problem. Driving up to Lookout Mountain with the extra weight of the documentary crew had drained my battery. I needed to recharge, and one of the reasons for stopping at the Incline Railway was that I knew there were two Level 2 public chargers in the parking lot. However, when we arrived, we quickly learned that *neither* of them worked. A company called BLINK owned them, and while it was a national company, it had recently been having some financial difficulties, I'd heard. Perhaps that's why the chargers didn't work, but, regardless, I was now in a bind.

I quickly did a search on the Recargo app and found that there was

Lookout Mountain Incline Railway from visitor platform. www.ridetheincline.com/ take-a-ride

one charger at Ruby Falls, which was a little farther up the mountain and on the other side of it. What luck! In order to save power, we slowly made our way back up the mountain and over to Ruby Falls.

Because it was the middle of summer and tourist season, it wasn't surprising that the Ruby Falls parking lot was completely full. We hunted around for the public charger and found it. It worked! But…. the owner of a huge SUV was parked in the parking slot, so I couldn't access the charger.

Herein lie two major drawbacks for consumers considering electric-vehicle ownership. There are not a large number of public chargers available in the first place—you are *forced* to plan ahead and to *know* where you can get power. *But* even if you do your homework, when you get to a public charger,

1-the charger may not work for a multitude of reasons (this has gotten better now), or

2- there may be some self-important SOB who decides to pull into the designated parking spots that electric vehicles use to recharge.

Both impediments can cost you a lot of time, and the latter one can *really* piss you off.

Since we had no choice, I maneuvered the LEAF as close to the rear of the SUV as I could. After a few tries, I finally got it close enough for the recharger cable to barely reach the connection to charge. And then we waited—I and the camera crew—for the owner of that SUV to return, and the camera crew would start rolling the cameras, and we would all confront the owner on why that SOB felt compelled to use the EV spot. But sadly, the owner didn't return during the one and a-half hours we

From left: George, Benswing, Rachel, Evan, and Jonathan take in the view at Lookout Mountain.

were charging there. Maybe the owner saw the cameras and decided that returning wasn't a good idea!

We had a late-afternoon event in Chattanooga, so we boogied on down the mountain. We were, again, a little late, so the group was waiting for us. They gave us some hearty cheers as we drove in, which felt good. Here we had some EV owners, a couple of press folks, a few interested consumers, and even some environmentalists. After Susan gave an interview, we all went inside and discussed the Tour and electric vehicles. Stuart even got to present his case for needed legislation to avoid environmental disaster because of global warming. Stuart showed compelling slides about carbon emissions and global-warming trends, I thought (even though I don't consider myself an environmentalist). His presentation, however, was too long, and his monotone voice made his speech as dry as a Sahara windstorm, but his heart was in the right place!

Stuart did cite some very interesting statistics:

➤ Earth had its fourth warmest year on record in 2013.

➤ ALL of the ten warmest years on record have occurred since 1998!*

➤ Climate scientists are 95 percent–100 percent sure that human activity (i.e., emission of greenhouse gases) is the dominant cause of dramatic warming.

➤ Impact:

  • Rising sea levels.

*One of Stuart's Global Warming slides showing
the impact of carbon emissions on temperatures.*

- Acidifying oceans.
- Melting glaciers.
- Intensifying heat waves, downpours, droughts, and wildfires.

This type of evidence certainly has to make you ask yourself: *"What if the scientists are right?"* What can we do or should we be doing?

It was a good event overall, and we felt good about our cause as we headed to the hotel that night.

*Data through 2011

CHAPTER 10

Day 7: Chattanooga to McMinnville

"Mayoral Mayhem"

"In Tune at the Commune"

Tuesday, July 10. A day forever etched in my memory bank—and probably everyone's that was on the Tour.

The drive from Chattanooga was around eighty-five miles and would take us through some nice backcountry between the mountains of Tennessee—past Signal Mountain and near towns with classic names such as Falling Water, Soddy Daisy, and Beersheba Springs. We also had extra bicycle and scooter riders today.

Included in our group, we'd have the pleasure of the mayor of McMinnville (our destination city) and three of his city managers join us on electric bicycles. Now one might assume that they're all outdoor enthusiasts happy to get outside for the day, right? Well, not exactly. It turns out that the mayor is retired, transforming his town "green," and is a big supporter of electric vehicles, so when he heard about our Tour he decided to join us and mandated that his managers ride, too!

We met them in the parking lot of our hotel in Chattanooga and talked about our Tour and our route for the day. They thought we were taking too much highway, so they gave us a new and more scenic route, which it is was! The focus turned to our vehicles, and the mayor was the first to try riding the electric bike. Well, most people don't realize that electric bikes have immediate torque, and he quickly took off and nearly crashed that

Herding cats along the back roads to McMinnville.

bike at full speed in the parking lot! Thank God we didn't kill the mayor of McMinnville that day!

I'm not sure how thrilled the three managers in their forties and fifties were to ride a bike eighty miles, but these guys were funny, and determined to have a good time. They were constantly cracking jokes or making fun of their riding mates. Even better, they were impressed with the electric bikes, which made it much easier for them.

Still, keeping the riders together on a two-lane road was sometimes a challenge because all had their own comfort level. Most important, the mayor and his three managers had fun; we had fun; and all made the trip safely through the rolling hills of Tennessee.

As we approached McMinnville, we learned that the mayor had arranged a special stop for us. The Cumberland Caverns are a National Natural Landmark with thirty-two miles of caves, making it one of the longest in the United States and the world. It's more widely known for another activity, however, something called Bluegrass Underground.

Bluegrass Underground is a hidden gem in mid-Tennessee. A PBS award–winning concert series, it features live entertainment for about 600 people—underground in a cave! The Volcano Room is 333 feet underground with a constant and pleasant year-round temperature of 54 degrees F. Small bluegrass bands and big-name country artists such as Lee Ann Womack and Chris Stapleton have graced the stage in the Volcano

*The documentary crew climbs out of our
crowded LEAF at the Cumberland Caverns.*

Room, and so did we!

We met our guide for our tour, and I learned that I had something in common with him: to take the entertainers down for their performance, he drove them down into the cave—with a Nissan LEAF! Why? Because the LEAF needs zero gas and emits zero pollution, so it's the perfect vehicle to use. And since all of our vehicles had zero emissions, he invited us all to drive down into the cave with him. We didn't have to be invited twice, so in we all went, one by one down the windy and narrow dirt-and-gravel road to the Volcano Room, past stalactites and stalagmite, past waterfalls and pools of water. The short drive was fascinating!

Of course, this was a relatively easy ride for the scooters and bicycles; as for *me*, not so much. You see, I was driving a *CAR* into a *CAVE*, and the entrance to get in was *extremely tight*. I mean, there could not have been more than two inches of clearance beyond my mirrors on either side. The car *just* fit, and only with some guidance from Stuart and the documentary crew did I get it in there. And I remember telling myself: "We are not going to tell John (Arnesen) or Nissan about this." And I never did—until now!

We all met up in the Volcano Room and admired the view, relaxed, and watched while Benswing and Dominique did a little square dance of sorts. We learned from our guide that one of the reasons that Bluegrass Underground is successful is that the acoustics of the Volcano Room are quite good as a venue for a concert. So, of course, we had to try it out by singing a song. No one could really come up with one we all would know, so we ended up singing "Amazing Grace" which, while maybe a bit somber, might just have been the right song to sing in the Volcano Room.

After a few pictures, it was time to head back up out of the cave. We thanked our guide for an amazing experience, which, no doubt, left a lasting impression on us all.

Driving into Cumberland Caverns was one of the unique experiences

A live performance in the Volcano Room.
(Photo courtesy of Cumberland Caverns.)

After getting in, I had to maneuver the LEAF down the
narrow road to the Volcano Room.

Group photo in the Volcano Room before we sing "Amazing Grace."

I've had in my life. I hope to return someday soon to listen to a concert in that same Volcano Room; I doubt I'll be able to drive another LEAF down there!

Once again, we were running late. We had an interview set up with WCPI, the local public radio station in McMinnville. So it was decided that Jonathan, Benswing, and I would drive on into town to do the interview. This was quite exciting because it was a live radio broadcast where we answered questions about the Tour's mission, details about the vehicles' capabilities, and our experience so far.

Meanwhile, the rest of our crew were met by the local police depart-

ment, this time on a more positive note! In fact, they were there to provide a police escort for the Ride the Future Tour and the Mayor into McMinnville. As the police cars rode into town with sirens blaring, the rest of our crew were greeted by a rather large group of townspeople who were waiting to start an evening of festivities in our honor.

In fact, the town ran with

Jonathan, Benswing, and I doing a live radio interview at WCPI.

Susan's event ideas and did a fabulous job with

- a small car show of electric and hybrid vehicles,
- a test drive for the electric bikes and scooters,
- a singing contest,
- a painted town mural of children's handprints with tips about being "green,"
- a few speeches about the mayor's day and the purpose of our Tour,
- an evening picnic and social time.

The highlight was the mayor presenting Susan with the key to the city!

The people of McMinnville could not have been more cordial. They even posted a "WELCOME RIDE THE FUTURE TOUR" greeting on the marquee of the local theatre for us!

It was a fun evening and an uplifting one. This is what we hoped the Tour would have been like in every city. As the evening wore down, I got one of the managers from the local power company to show me to its facility, because it had the only recharger in town. Just as I plugged in, another nasty rainstorm broke, and I left the LEAF hoping that the power didn't go out and that I wouldn't return to a dead battery in the morning.

I met with the group as the event closed and was greeted by a friendly new face. Susan had arranged for us to stay that evening with a local group called Isha, and some Isha members had made sure to introduce themselves to us during the event.

The Isha Foundation is a nonprofit spiritual organization founded in 1992 by Sadhguru Jaggi Vasudev. (Don't worry; I can't pronounce it, either.) It is based at the Isha Yoga Center near Coimbatore, India, and at the Isha Institute of Inner Sciences at McMinnville. The foundation offers yoga programs under the name Isha Yoga and its slogan is "the silent

revolution for self-realization." It has over five million volunteers and works with international bodies like the Economic and Social Council of the United Nations.

We put all of the electric vehicles inside the nearby town hall, left all its electric sockets full of charge cords of batteries to charge overnight, and stepped into a van with the Isha driver to take us to the foundation's community. It was now late evening; we were all tired; and I swear the drive took us halfway back to Chattanooga! The Isha Foundation owns 1,200 acres of land on the Cumberland Plateau, and this place is in the middle of nowhere—probably by design.

When we arrived, we were offered some food. We hadn't really eaten much at the event in town, and since we didn't want to be rude, we took it. It was a form of soup, but I have absolutely no idea what was in it, veggies of some sort, I guess. We got that down and decided that we were all "full" now.

Isha has residences for their members, as well as some apartment-type facilities for visitors; I guess they do corporate seminars for interested parties at times. There were three or four of us to a room, and the facilities were pretty standard; nothing really stood out—except for one thing. There were no locks on the doors. None.

Mind you, the day before, the Tour group had joked among one another about the possibility of being taken to another "Jonestown" and never seen and heard from again (i.e., The Lost Electric Parade). The joking was a nervous laugh. Yes, everyone we'd met had been very friendly, but the fact there were no locks on the doors felt a little creepy. I'm sure it was somehow connected to "self-realization," but I went to sleep that night, just hoping I'd wake up in the same bed in the morning!

Thursday morning, July 11

The Isha Foundation is based on a program called "Inner Engineering," which revolves around meditation. And being our hosts, the foundation's members wanted to share their philosophy, so they invited us to their morning ritual. We walked over to their domed building, where everyone in the community met. Yet Mahima Hall, the domed building, is not just any meditation space; it's a 39,000-square-foot dome, reportedly the largest domed meeting room in the world!

Once inside, we were asked to take our shoes and socks off before congregating in the front of the room, where a senior member would take us through the ceremony. A picture of our "supreme leader" was propped on a stand either for us to honor him or so he could look over us, I'm not sure which; I couldn't tell if the eyes in the picture were moving or not.

We went through a series of "repeat after me" phrases (spiritual wisdom) followed by some meditation sequences, highlighted by loud *huuummmmms* (i.e., chants), which, with your eyes closed, were designed to relax your body and mind (and your wallet slip out?). I'm not really sure how many of these we did, but, by the end, I was beginning to get a headache from trying to be in tune with the commune and all the long *huuummmms*.

It was a "different" experience.

Isha's Mahima Hall where we gathered to meditate;
note the leader's picture in the center
http://isha.sadhguru.org/us-en/isha-usa

After the ceremony, I was expecting Kool-Aid, but, fortunately, there wasn't any. Seriously, the people from Isha were very hospitable to us, and I appreciated it. I may not become a member soon, but obviously there are a lot of people who believe that meditation and yoga play a key role in their physical and mental well-being, so perhaps it is my own stereotyping that forms an image of a cult, and I am actually the one missing out.

CHAPTER 11

Day 8: McMinnville to Nashville

"There's Silver in Them Thar Hills."

"Rainy Days on Mondays, Tuesdays, Wednesdays. . . Everydays Always Get Me Down."

By the time we finally got back to McMinnville and got all of our equipment out of the town hall, it was already 10:30 a.m. With another city event at Centennial Park in Nashville in late afternoon, we needed to get moving.

As we left town, the two-lane road widened to a four-lane highway through some rolling hills. The ride would not be too bad at ninety miles, but traffic would be heavier. I was supporting BenHop on the bike with batteries this day, and we'd left a little earlier than the rest. But soon the scooters caught up, and we were cruising along at our normal 25 mph or so. One by one the scooters began passing me, Susan followed by Dominique (music headphones and all) and Rachel.

Susan was the master at multitasking while she rode the scooter. She was always trying to arrange another future hotel, or talking to a city about getting an event together for us, or talking to her property managers in New York. This day was no different, and as she slowly cruised by me in her pink helmet, I could see that she was talking on the phone. She had gotten ahead of me by maybe 100 yards when, all of a sudden, I saw her purse drop from her scooter *ker-splatt* onto the road. From here, it looked like a miniexplosion, with coins rolling down the highway, money and receipts flying in the air, credit cards slicing through the air like little Frisbees. Everybody, including me, immediately slowed, stopped, and jumped out to help. But recall that we were on a four-lane highway with traffic rolling at maximum speed and coming over a hill at us. So here's the visual: you have a group of five–six people dashing out into the middle of the highway to collect a purse, pick up coins, and chase dollars, receipts, and credit cards being blown all over the road as cars and big trucks *swoosh* by at 70 mph! I'm sure we got most of her purse contents, but no doubt there remains some silver in them thar hills!

Ben Hopkins, meanwhile, had continued his long biking trek ahead of us, not really aware of the chaos going on behind him. It was not unusual for Ben to get into his own "zone" while riding. During our journey, he never listened to music or an educational audio tape or anything else while riding. He just biked—one pedal at a time, constantly moving to extend the life of the bike battery. Incredible really. Maybe he had an internal jukebox or was solving world peace in his head, I don't know, but the man was a beast on the bike.

On this day, however, Ben was having issues with the bike. He first got a flat tire, which in itself was not that unusual; he'd had a couple of them previously that he'd just fixed. But today's problem also involved something with the spokes, and this delayed the whole team because it involved the supply truck and tools.

When we got rolling again, it was time for a very late lunch, so we stopped at a Texas Roadhouse for a bite to eat. We discussed the fact that we were running late again, and we decided that, with some help, Sean would take the supply truck to the park and set up our banner and display for the "green" event there.

While Sean headed out to the park in Nashville, we also changed our course. Instead of going to my house, we would head directly into Nashville to get to the event as soon as possible. But as we neared Nashville, once again our old nemesis struck—we hit more rain. Not only did we get wet, but Sean advised us that the city had been hit with a nasty rainstorm that had flooded a lot of Centennial Park. There was no one there, and there would be no event.

BenHop makes repairs to the wheel and tire as we wait.

Undeterred, we decided to go into Nashville after stopping at my house and recharging. There were always crowds downtown, so we would do some filming there.

We arrived downtown and made our way down Broadway, where all the honky-tonks are, and pulled into a circle on 1st street. BenHop was continuing to have issues with his bike wheel, so he and Sean decided they needed to rebuild the wheel. They found new spokes and tools and began to repair the bike just as it was getting dark. They finished the job by cell phone light and flashlight.

While we were waiting on the bike repairs, Susan noticed that a Channel 2 News reporter and truck had shown up right next to us and was doing a report on river activities or the fantastic fun the pedal bar bikers have or something simple like that. (If you haven't heard of pedal bar bikes, check this out: www.nashvillepedaltavern.com.)

Susan saw this as an opportunity and soon was engaged with the reporter, who subsequently interviewed her about our adventures and the Ride the Future Tour. I don't know when it aired, but once again our cause had picked up some notable exposure!

With repairs and interview completed, we all saddled up for a parade through downtown Nashville. We rode down Broadway honking our horns and screaming "Electric Power!" and "Ride the Future Tour!" or some such silliness with our cameras rolling. We didn't get much attention, but it was still fun. We did get the attention of the Nashville police, who stopped us and told us that Evan couldn't sit on the windowsill and film. Party poopers!

Once our honking had grown old, the group split up. Part of the group, like BenHop, would go with me to my house for the night to get rest. Part of the group was up for partying at the honky-tonks and would show up

Sean and BenHop proudly show their finished product after rebuilding the bike's rear wheel.

Susan gives an interview to Channel 2 News in Nashville while the documentary crew films it.

later. And Benswing, well, he was off to a swing dance party somewhere. Susan went to visit family.

One by one, most made it to my house for the night, crashing in my bed, on my sofa, in my upstairs bedrooms in a bed or on sleeping bags, basically wherever they found a spot. It had been another long day, but we were still moving.

CHAPTER 12

Day 9: Nashville to Natchez Trace State Park

"The Blair Witch Car Switch Project"

Evan arrived at my house around 8:00 on this Friday morning; he was reportedly dropped off by a pretty, young female. The documentary crew had had a good time the night before. As the Tour left Nashville, the documentary crew stayed behind to "download a lot of film they had" and to pick up a rental van, which would give them more room and freedom to film. Near the end of the Tour, however, they admitted that they were all hungover and needed more sleep time—and that George had puked in my back yard. Thanks, George.

I had to hustle everybody over to the nearby Nissan corporate office for a 10:00 a.m. meeting out in front of its national office building. Nissan had told employees that we would be outside, and a number of employees came out to greet us and talk about our trip, including the Internal Communications Department. John Arnesen was there, and we discussed the fact that I had already broken the old Guinness record for distance in an electric vehicle but I was charging ahead for more. John and I gave interviews to the PR folks, and chatted for about an hour with employees.

When the Nissan activity was over, we went back to my house to prepare for the day's trip, which would be a long one, about 100 miles. Everyone was ready to go pretty quickly—except me. I'd lost my house keys. I looked. Others looked. For over an *hour* we looked. I finally found

A group picture in front of Nissan's corporate office, along with a few friends.

them—in a second level of my pants pocket. They'd been on me the entire time, and we were delayed an hour. Ugh!

We left the documentary crew at my house and started our trek down the Natchez Trace, a part of the National Park Service and a ridgeline trail that extends 444 miles from Nashville, Tennessee, to Natchez, Mississippi. It's a relaxing but curvy two-lane road that winds through miles and miles of forest and farmland while passing historic sites and natural wonders like waterfalls and scenic views. There's not a lot of traffic, so it's a popular ride for motorcyclists and cyclists. It was perfect for the Tour.

Once again, with the long-distance ride, Benswing and I went ahead so we could recharge in a little town called Linden. We enjoyed many Natchez Trace scenic and historical sites along the way, but eventually, by late afternoon, we exited the Trace and made our way to the Commodore Hotel, where the charger was located. We plugged in and just happened to run into one of the owners of the Commodore, Kathy Dumont.

Kathy, we learned, was committed to sustainability and was hoping the chargers would bring in new business to the Commodore. She and her husband, Michael, were originally from Rhode Island but moved to a farm in the Linden area. They then bought and restored the 1939 hotel to a quaint twenty-two-room stopover in its original era style. They added a café, as well, which features good food and live music.

After some fun conversation and a good meal, we headed on to meet with the riders who were getting near Natchez State Park. By the time we got to the park, however, it was dark. Benswing, Rachel (my rider for the

*Benswing and Rachel at base of Windsong Hollow Bridge on the
Natchez Trace; Rachel and I at the top.*
www.natcheztracetravel.com

day), and I started driving into the park, looking for signs for the camp-ground. We had been driving for maybe twenty minutes or more when we caught a glimpse of taillights ahead of us, and, as luck would have it, they belonged to BenHop's electric bicycle and the scooters.

We spent a few minutes catching up and then continued on. We drove, and we drove, and we *drove* to find this damn campground. I felt bad for BenHop because he had to have biked 110 miles by now and he had no headlight, so he was relying on others behind him for light. Finally, the group stopped ahead of us to discuss whether we were still going in the right direction. It was pitch black, and we couldn't see anything other than where headlights shone.

While we were sitting there, a vehicle we didn't recognize pulled up behind us and stopped about seventy-five yards away. The driver (whom

*Benswing and Kathy Dumont lean on the
charging LEAF at the Commodore Hotel.*
www.commodorehotellindentn.com

we couldn't see) turned off the vehicle's headlights. Rachel and I started to get concerned because the vehicle just sat there. The driver obviously didn't want to go around us—for some reason. We worried that the vehicle's driver and passengers might be folks planning to rob us. They just sat there. We nearly panicked because this was very strange, with total darkness around us. Finally, someone got out of the car and started to approach us from behind. I locked the doors and told Rachel that if that person tried to rob us, I was going to floor the accelerator. We waited, both of us looking in the mirror to see who was behind us. Suddenly, Rachel shouted: "It's Jonathan!" (our documentary crew director who was with George and Evan still in the car.) Jerks! They were filming us the whole time to watch our reaction.

We all got a good laugh out of that, but we still weren't at the campground. As it turns out, the roads (not shown on most GPS devices or maps) form a "V" shape inside the state park and we had started from one end and driven to the other end of the "V" shape to get to the campground. With little choice, we kept going in the dark and finally found the campground around 11:00 p.m.

I drove the LEAF downhill into the campsite and pulled up to the power outlet. This was the first time that I had to use one of the two supplemental cables that AeroVironment had given me for our trip. The outlet was an RV outlet of 220 volts, which I connected to a portable charger to charge the car.

Struggling in the dark, I did get everything connected and, voila, the charge indicator on the LEAF lit up. Another small milestone had been achieved.

It had already been a very long day, and the thought of pitching tents in a campground (Susan's idea for bonding, I guess) in the dark was not very exciting.

We pulled all the tents and sleeping bags that Susan had bought for us out of the supply truck and began to try to set the tents up. Fortunately, Sean, the smart one, helped out, and we got them all up. Meanwhile, someone had started a fire, and we cooked some food and roasted marshmallows that night. Those that hadn't had S'mores before got introduced. We were all exhausted by the time we went to bed around 1:00 a.m., but Susan's camping idea was a great bonding experience, and we actually had a fun time with it.

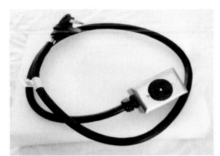 

Supplemental RV outlet cord (l) from AeroVironment
plugged into a portable charger (r).

Our first camping adventure as a group in Natchez Trace State Park.
http://tnstateparks.com/parks/about/natchez-trace

CHAPTER 13

Day 10: Natchez Trace State Park to Bolivar

"Men are Waffles; Women are Spaghetti."

"Setting the Nightly Alarm to Wake Up to a New—Battery?"

A good friend recently told me:

Men are Waffles.

When it comes to incoming information, they like to deal with it directly, process it, make a decision on it, file it away in compartmentalized files, and forget about it, never dealing with it again (or at least they know where it is for future reference).

On the other hand,

Women are Spaghetti.

When it comes to information, they take it in, process it, may or may not deal with it immediately, let it wind around through their brain forever. Even when a decision is made, previously processed info can be brought back to the surface. It may be a day ago, a year ago, or ten years ago, but all of it can be retrieved and assimilated with current info for formulating a new result or connection.

The revelation here is that this contrast between men and women explains

1) why women don't understand men's directness and their having the ability to "forget" the past, while

2) men cannot fathom women taking current incidents and connecting them to past history.

This contrast clearly shows up in relationship issues, and, I believe, it applies in general.

I *am* a waffle. I like to plan, organize, and make decisions once I have all the relevant information and to be happy with those decisions going forward. It may take me a while to make them, but I'll be happy with them and never look back.

Stuart is my friend. He is also a waffle. He likes things very organized and efficient. He *needs* eight hours of sleep every night, and he's a creature of habit; he does the same thing in the same way every day. It's good—and it works for him. But it doesn't always work for others.

Despite our campfire Kumbaya the night before, there had been rising tension in the group. Rain was a factor not only with minimizing our public events but also psychologically. There wasn't much we could do about the rain. But the bigger impact it had was our delays before we arrived at our destinations.

When we arrived late to our destination for the day, we all went to bed late, got up late, and subsequently delayed our departure from the time we had all agreed to leave, which was supposed to be 7:00 a.m. In fact, lately, our departures had been more like 9:30 a.m. or even 10:00 a.m.

We already had full days, which basically consisted of

- packing suitcases and equipment and placing them into the supply truck;

- eating breakfast and having route/travel/passenger discussions for the day;

- traveling 40–50 percent of daily distance;

- eating lunch and having break time (recharging took two–three hours for Benswing and me);

- traveling the remaining 50–60 percent of daily distance;
- meeting with consumers and/or press events;
- holding a group dinner;
- unloading equipment and suitcases from the supply truck;
- charging vehicle batteries;
- answering e-mails, doing social media postings, discussing issues, and more;
- going to bed;
- repeating the next day what we did the day before.

Stuart had grown frustrated because he has a regular routine of getting eight hours of sleep. However, when we arrived late at our public events and if we planned to leave at 7:00 the next morning, he didn't get eight hours. But then, when we didn't leave until 9:30 or 10, he got upset. His frustration first boiled over in Chattanooga in front of our cameras.

I was also frustrated because we had been late to some of the scheduled public events. And others were frustrated because the pace had been hectic lately; there was no room to relax if we were to stay on the forty-four-day schedule.

I think Susan had been hearing some of this frustration, and she elected to call a morning meeting. We all huddled up at a picnic table under a shelter and had an open discussion. It was sometimes contentious; not only did we have vehicle-speed and charging differences, we had other differences as well, not the least of which was a large age range and lifestyle differences among our group.

It was a good interaction, and a few things came out that I didn't know. One obstacle that came out was that the girls, who charged all the scooter and bike batteries, never had enough outlets to charge all the batteries in their room, so they had to set alarms for every three hours to get up and change the chargers to a different battery. They weren't getting a full night of rest, either. This was one of the reasons they were sluggish on being ready early the next morning.

There were concerns about the pace as well. One idea floated was to have an occasional day just to relax and breathe along the route. BenHop made the comment, "Of course it's difficult; we're trying to set a Guinness World Record, so it's supposed to be difficult." Susan, to her credit, stated she wasn't open to extending the length of the tour more days. Her concern was that she had made commitments to a number of cities along our route and that if one changed, they *all* had to be changed.

I pushed for leaving early every day for three main reasons:

> ➢ To make sure we arrived at our destinations on time for events.

> ➢ In future segments like the desert, we would want to minimize our afternoon travel time.

> ➢ If we got to bed earlier, we would all be better-rested and on time.

In the end, we all pretty much agreed to leave earlier, though, at times, BenHop on the bike would have to leave even earlier since he had the slowest-paced vehicle. For him to do that, it meant that either the supply truck or I myself would have to stay with him to provide a regular supply of batteries during the day.

Overall, it was a good meeting that we needed. It relieved some tension. Stuart was still frustrated by the fact he had few opportunities to speak in front of groups about global warming, which he had agreed to with Susan. Sean was frustrated with always being the supply-truck driver. By the end of the discussion, we knew everybody was frustrated with something, but we basically sucked it up for the good of the Tour. We had to if we were to be successful.

The group spent the rest of the morning and early afternoon relaxing. Several people went swimming in the lake; a few had rafts; some swam; and BenHop lounged around like a polar bear on an iceberg. The heat and humidity were pretty brutal in July, so the water was a welcome relief. I am, however, still kicking myself for not having taken a picture of Stuart on this day. Stuart, wearing his wide-brim, safari-type hat in his bright blue Speedo was priceless!

We finally got kicked out of the campsite around noon by the Park Rangers because someone else had our site for that night. So we packed up to head out.

Once loaded, I backed uphill out of the entry to the campsite and inadvertently hit the wooden site marker. It left a small dent and scratch in the rear quarter panel of the LEAF, which, since it was Nissan's car, bummed me out. But it was small, and, fortunately, it would turn out to be the only

Sean and several of the team enjoy the cool water in Natchez Trace State Park.

mark I got on the car for the entire 5,100 mile trip. More important, I didn't have to pay for repair out of my deductible!

We had the luxury of relaxing and leaving late today because our destination was only about sixty miles away. And after lunch in the small town of Henderson, we made our way to Bolivar, Tennessee, and the Bolivar Inn.

The owner of the motel was very nice and supportive. He helped us find power for the LEAF by plugging it into a 220v outlet between the vending machines.

It worked! The LEAF would be juiced up in the morning once again.

After a meal at one of the local eating establishments, we packed it in for the night. Though Bolivar is "the Quintessential Town in West Tennessee," the only "quintessential" thing we needed from the town on this night was a good sleep.

Our motel in Bolivar had just the thing for "Green Lightning" to cool off with—vending "juice"!

CHAPTER 14

Day 11: Bolivar to Memphis

"Roach Coach Rodeo."

The group got an early start this morning; a good sign after our meeting the day before. It was about a seventy-mile segment, so not bad. Our trip would take us to Shelby Farms Park, where Benswing and I would recharge. There was also some type of event there in the afternoon before we finished the day in Memphis.

Benswing and I went ahead of the group and arrived midmorning. We learned that sustainability is one of the park's core values. Managed by a nonprofit conservancy, Shelby Farms Park encompasses some 4,500 acres of green space with a master plan to make it a leading example of parks for the twenty-first century. Activities in the park include

A buffalo herd	Laser Tag and Paintball	Disc Golf
Bicycling	Boats	A playground
Eleven-mile trail to Memphis	A BMX bike race track	Trails
Lakes	100 acres of off-leash pet area	A walk bridge
A treetop adventure	Horse stables	

To be a twenty-first-century park, Shelby Farms Park has to have electric-vehicle chargers, right? They do. Ten of them! The Smart Modal Area Recharge Terminal (SMART) station, which is located in the parking lot of the Shelby Farms Park Visitor Center, captures solar energy to offset the energy used to charge plug-in vehicles. It also collects data from users that will be used to track infrastructure trends and forecast the design and

*On the road again in Bolivar and making our way
to see the buffalo in Shelby Farms Park.
www.shelbyfarmspark.org*

impact on the electric-grid network. The SMART station was impressive! And a real indication of what the future could look like.

It didn't take us long to recharge, so, by noon, we moved our vehicles over to the parking lot, where our event would take place. We learned that the park had fostered a regular routine for Sundays at noon, when numerous catering trucks in the area would gather. As a result, a lot of park patrons and visitors came to the parking lot to feast on their personal savory selection. After the rest of our Tour arrived, we spent a couple of hours talking to a regular stream of visitors who wanted to know about our cause and/or take a ride on an electric bike or scooter.

While at the park, the Tour did take the time to enjoy some of what the park has to offer. But around 3:00 p.m., BenHop had to leave for an interview in Memphis, so the group split up. Benswing and I headed off to visit Graceland, since we had some free time. I had never been to Graceland. I'm not a huge Elvis fan, but I certainly appreciate his talent and his benevolent nature. The grounds of Graceland are an amazing story of his life, and our visit there was time well spent. I even had a funny thought while leaving Graceland: "Do you think Elvis would have traded in a Cadillac for a LEAF?"

Benswing points to where we are in our journey on the LEAF as we charge at the SMART station.

As I made my way toward the hotel, I realized I was getting low on power. so I looked up on Recargo the closest charging station near me. That turned out to be Christian Brothers University, and I made my way there. When I arrived, I didn't know where the charger was. Naturally, I asked a security officer who was driving by in a golf cart for directions. He then gave me the third degree about what I was doing and why I needed the charging station. I told him it was showing up on Recargo as an available public charger and I was low on power. He wasn't convinced I was in need and cited it was a private university.

It took some convincing and a promise that the cost to charge was minimal before he would agree to show me where the charger was. A campus administrator gave me an assist, so I let him display his *Get BUC Wild* shirt for a pic. Go Buccaneers!

Some of the group decided to go downtown to Beale Street in Memphis to eat and enjoy some blues music. I opted out this time. But I was glad we had made Memphis. So far, everything had gone to plan. Just over the Mississippi River, which we would cross in the morning, was a new frontier for me. The cities with public chargers would not be so close. And not every area had an RV campground. In fact, in a number of areas going forward there were only tent campgrounds or undeveloped (no facilities like power or water) campgrounds. And one of the "holes" in my plan for power lay just ahead.

BenHop makes his food choice for lunch from a wide choice of catering trucks.

The group takes in some of the Shelby Farms Park activities.

www.graceland.com

Recharging at Christian Brothers University in Memphis.
www.cbu.edu

CHAPTER 15

Day 12: Memphis, Tn. to Brinkley, Ar.

"We'll Cross that Bridge When We Come To It. Oh, We're There?"

"Station to Station Power Elation."

On Monday, July 15, we motored into downtown Memphis and assembled at a park by the Mississippi River to plan our crossing of the bridge. Two months ago, I'd identified this location as a problem child. In the entire Memphis area, there are only two ways to cross the Mississippi River—and both of them via interstates. That was problematic for two reasons:

> ➤ Speed: at our 25 mph, traffic would be on us immediately and pose a serious risk of accident and injuries.

> ➤ Legality: most interstates do not allow vehicles to travel below a certain speed, and while we could ask for a highway patrol escort, the patrol might deny it.

Neither option was a good one. And, to make it more interesting, Jonathan wanted our crossing all on film!

With some nervousness, we decided to try crossing undetected as fast as we could. Our strategy would be for me to lead the group because Jonathan had wanted to film it that way. Sean would protect our rear flank

Susan finds a little shelter from the hot sun as we prepare to cross the Mississippi River.

with the supply truck and its emergency flashers. The documentary crew would also be involved by starting out beside Sean, thus blocking traffic and making sure no cars cut back into Sean's lane after passing him and running right into BenHop and the girls. With traffic speeding by us at 70 mph as we came off the on-ramp, our strategy was risky, but it was the best we could do.

We briefed everybody on the plan and headed for the on-ramp. When I could see everyone right behind me, I pushed forward onto the ramp. As I saw a traffic break, I accelerated out into the right lane with my emergency flashers going. I wanted anyone coming up behind us in the right lane to move over as we merged on. All of us hit our throttles as hard as possible, including BenHop on the bike who was at most risk. Sean cut over as quickly as he could to provide cover for the lane, and Jonathan followed suit. We were on our way.

The bridge was a massive one, with a long span, of course. Jonathan cut in pretty quickly into the left lane, with Evan already hanging halfway out the window to film. Traffic immediately began closing in and stacking up behind us. I was scared that we might cause or be in a bad accident. At minimum, we'd get angry drivers and horns honking, but, surprisingly, there weren't any. Why? I looked in my rear-view mirrors to find that a couple of eighteen-wheelers had pulled up behind us first, and when their drivers saw that Jonathan was filming and what we were trying to do, they arranged their rigs side by side and blocked traffic for us! We couldn't have asked for anything better! They gave us shelter all the way across the bridge. As we exited at the very first exit we could, we gave them a hearty wave. They had done us a huge favor. Jonathan even got his filming from behind, from the side, and from the front—just as he'd wanted. Our nerves were still a bit on edge, but we had made it safely, and smiles were all around. We moved on to East St. Louis and stopped at a grocery store for a drink.

Jonathan also helped us in another way. He had hired a PR and marketing person for us so that Susan didn't have to do it all. Susan had done and was doing so much during this whole adventure, so Jonathan's help was a big step for her. She didn't have the time to coordinate with any PR person. She spent two hours on the phone with Morgan (Jonathan's PR lady), briefing her on status and needs while we had lunch.

Arkansas was not only new topography with a lot more low-level water bogs and swamps (versus the mountains, trees, and hills of the East), but the population was also more spread out. The back highways here were more desolate and peaceful, and thus enjoyable from our standpoint.

But fewer people meant fewer electric vehicles. One thing you learn quickly if you're an electric-vehicle owner is that power is king. It forces you to plan and to know where you can get your vehicle recharged along your route if recharging is needed. If you don't have power, you can't get a gasoline can to fill up your vehicle; you just don't move. At all. The ironic thing is that power is all around us. Every home, business, or facility has power of some sort coming into it; the question is whether you can tap into it and whether it's in a form that you can utilize.

We were headed to the little town of Brinkley, Arkansas. According to the recharging websites, there were no public charging stations anywhere nearby. Period. Moreover, there was no RV park in the area to tap into. I was stumped on what to do. The only other alternative was to charge with 110v charger, and that could take as many as fourteen hours to fully recharge the LEAF. Ugh, ugh. With the possibility I wouldn't have a full charge to start the next day's drive to Little Rock, I could fall behind the rest of the group.

Susan had spoken to the motel owner at my request, and he had told her that he had 220v power that we could use. Susan was adamant that we had to have it. He understood. Nevertheless, as we pulled into the Days Inn, I had my doubts. The owner wasn't a native, so his English wasn't easy to understand, but he told me that we could unplug the air conditioner in the room, which ran on 220 volts of power. News to me, but okay.

The next questions, however, were harder, and those were: what type of outlet was it?; would it fit either of the AeroVironment cords that I had in the car that hooked up the portable charger? If not, I would be forced to charge with 110v charger all night long.

In looking at the outlet, I could see that there was no way the RV plug would work. So I dug out the other cord, which I hadn't used yet. Remembering all the NEMA chart possibilities for 220v outlets, I was thinking: "What are the chances?" I tried plugging in the cord to the wall outlet. To my shock, the plug fit!

Before long, I had the LEAF charging again. Well, not before I accidentally bent the window screen, trying to get the plug and short cord to reach the outlet in my room—I believe that cost me $40. But, most important, the plug had worked. And if it worked at this motel, hopefully it would work at the next motel, and if so, *I had the bridge for my path to the West Coast. This was huge news!* As I would soon learn, every motel that has an in-room air conditioner in the United States uses the exact same outlet nationwide. With this revelation, I just might be able to make it all the way to SF. Who would have thunk? And the revelation came out of nowhere—in Brinkley, Arkansas.

Even with my window ajar for the cord, I got a restful sleep in Brinkley that night.

For the first time all trip, the LEAF gets charged from my room A/C; it would not be the last!

CHAPTER 16

Day 13: Brinkley to Little Rock

"They're Gonna Have a Cow Over This."

We got another good start Tuesday, and we cruised along the backroads of central Arkansas. With little traffic, we could relax and enjoy the scenery, which was much more bayou-like than I had expected. There were a lot of algae-topped ponds and rivers, which went along with the humidity, I guess. Every once in a while, there would be something unusual to pique our interest. One such oddity was an enormous train drawbridge at a river crossing. We stopped to check it out, but it appeared it hadn't been used in years. We marveled at how much steel and effort must have gone into making it. But, sadly, it's obsolete now.

Another oddity that popped up was the number of crop-dusting planes that were operating as we drove through. You would think that using airplanes would be an expensive approach to crop treatment, but apparently not. In fact, the documentary crew took a side trip to interview a crop-dusting pilot and got some amazing video!

For lunch, we stopped in a sleepy, former railroad hub town called Lonoke, population: 4,200. Lonoke got its name for the lone red oak tree that stood for years between two prairies. I wish I had a good story to tell about Lonoke, but sometimes it's better to let sleeping towns lie. And so I shall.

Our destination for the day was about seventy miles past more pine

Typical scenery in central Arkansas between farms and a massive vertical train drawbridge.

Crop dusting in Central Arkansas.

Lonoke's Chamber of Commerce stands in the old train depot near the Chinese restaurant.

trees and swampy land amid humid but sunny weather. We rolled into Little Rock in early afternoon and drove to another Susan target, this one called the Heifer Village. I had never heard of Heifer International, but I was up for a tour. Surprisingly, I learned a lot. Heifer International's mission is to work with communities to end world hunger and poverty and to care for the Earth. Its basic philosophy is "teach a man to fish," and it's

*Highway workers stop to look as our little electric parade
drives through a construction zone.*

been carrying out its mission for over seventy years!

Heifer links communities and helps bring sustainable agriculture and commerce to areas with long histories of poverty. It donates animals and services to those areas which provide partners with both food and reliable income. Agricultural products like milk, eggs, and honey can be traded or sold at market.

The goal of every Heifer project is to help families achieve self-reliance. Heifer does this by providing these families the tools they need to sustain themselves. The core of its model is "Passing on the Gift," which means that families share training they receive and then pass on the first female offspring of their livestock to another family. This extends the impact of the original gift, allowing once-impoverished families to become donors and full participants in improving their communities. Heifer empowers families to turn hunger and poverty into hope and prosperity.

Heifer relies on donors to fund a variety of projects, with investment ranging from $10 to $5,000. What's intriguing is that, as a donor, you can be quite specific on how the money is used, including

Heifer International's Headquarters at Heifer Village.

sending a girl to school, installing irrigation pumps, launching a small business, buying animals such as goats, ducks, honeybees, chicks, water buffalo.

You can also buy a whole animal or a piece of one. For example, if you want to buy a heifer (baby cow), a share is $50, and a whole one is $500.

Animal donations can provide families with a hand, increasing access to medicine, school, food, and a sustainable livelihood. If you'd like more info, please go to my website or to www.heifer.org/gift-catalog/animals-nutrition

Heifer impressed me as a very down-to-earth organization, and I love their "play it forward" approach to training and breeding animals for sustained success. So I'm a fan now—and a donor!

After our tour, I walked back to the LEAF that I had left charging at a public charger in front of, get this, President Bill Clinton's Presidential Library. I had hoped to take a visit through because it is supposed to be quite interesting. But the library was closing just as I got there, so I had to settle for a magnet to add to my collection.

I do have the telephone number, and I was there, but, to paraphrase Bill Clinton himself:

"No, I did not have textual relations with this establishment!"

Well, I guess that's not exactly what he said, nor did I get to see the blue dress. So I didn't get a happy ending.

Disappointed, I made my way over to the hotel. There were no Tour plans for the evening, so I had dinner with one of my cousins, who lives in the Little Rock area.

We had a good time talking about the Tour and catching up. Always great to get the latest family updates, especially when "on the road."

Our guide, Lori, explains some of Heifer's current projects; an office wall mural.

Recharging at the Clinton Library facilities. www.clintonlibrary.gov

CHAPTER 17

Day 14: Little Rock to Oden

"Ode to Oden."

At the two-week point in our trip, things seemed to be going okay. We were still on track, and, for the most part, we were getting along with one another.

We were headed to our southernmost point today, a leg into the more remote areas of Arkansas. The distance was about 100 miles, which meant a pit stop in Hot Springs for Benswing and me to recharge. We left early and arrived at Riser Nissan around 10:00 a.m. We spoke with several employees and managers there, who were very cordial and made us welcome during our stay.

We left around noon and went into downtown Hot Springs and the Bathhouse Row District. Though still an active tourist attraction today, it was once one of the premier vacation spots in the country. Designated a National Historic Landmark in 1987, the Grand Promenade consists of eight bathhouse buildings built between 1892 and 1923. With beautiful architecture, the houses are quite large, with the biggest (Fordyce) having 28,000 square feet of space, three main floors, a basement, and two courtyards!

The bathhouse spas were built around forty-seven natural hot springs along Hot Spring Creek. Highly regarded as having therapeutic benefits but without the sulfuric smell that comes with most mineral springs, the spas' decline came during the mid-twentieth century because advancements in medicine had made bathing in natural hot springs less believable

Group photo with the sales manager at Riser Nissan in Hot Springs.

as a remedy for illness.

We cruised the promenade area, and Jonathan shot some film of us before we stopped for lunch at Granny's Kitchen for some home cooking. After some good food and yuks, we were all ready to relax in one of the spas, but, alas, relaxing there wasn't in Susan's schedule, so we moved on.

It was still hot and muggy (it had been since we started this journey) as our crew made its way toward Oden. Fun can happen anywhere. So when some of us spotted some kids swinging on a rope and dropping into a small river along our route, they decided to join in. Benswing, Sean, and Stuart all got in on the fun. I was supporting BenHop with bike batteries, so I took a few pics and went looking for BenHop.

I found him, sacked out tired and leaning on an old building. Well, not really, but we staged a photo to make it look that way just for kicks. Actually, he was doing what he was always doing when I found him—pedaling.

www.nps.gov/hosp/learn/historyculture/bathhouse-row-today.htm

Benswing and Sean try out a local tire swing over a creek in Arkansas.

Okay, okay, we actually staged this photo at the time, but it looks good, doesn't it?

We would stop for the night at Riverview Cabins, a camping location in Oden, Arkansas. I was nervous about this spot because I couldn't find much about it, especially regarding its having 220v power. Susan had assured me that the owners had promised to have power for me, so I was going along with it—not that I really had a choice; there was nothing around that I could see for an alternative!

We made it to the campground in late afternoon and met the owners, Ron and Sue Spurling, and their family. They were very intrigued by our visit, and Ron had even wired a special 220-volt outlet for me to have power for the LEAF. It worked. Once again, the hospitality we met along this journey continued to amaze us.

We had a fantastic time at Riverview! Ron and Sue and their daughters opened the food trailers to fix us some delicious food and ice cream. Afterward, all of us on the Tour kind of did our own thing, taking advantage of what the park had to offer. I ended up playing horseshoes with BenHop, Jonathan, and the crew. It was another successful day during our Tour.

After it got dark, we all crowded into one of the screened bungalows and listened to Dominique sing some of her songs and play her guitar. Dominique enjoys singing and hopes to become a solo artist. This was an opportunity for her to share her passion. She was often listening to,

Rachel enjoys dinner with the campground girls, and our Tour dines in a screened bungalow.

singing, or writing songs during the Tour, and on this night, she had the stage. We also found out that BenHop is a closet folk singer! And he's pretty good, especially when it comes to Bob Dylan (his favorite) songs. It was a mellow and relaxing evening of music with the Tour, the owners' family, and other campers in the park. Everyone, especially the kids, enjoyed our private concert.

Ron and Sue Spurling.

In the morning, we thanked our hosts, Ron and Sue, for their hospitality. They'd done a great job in accommodating our needs, and we'd enjoyed getting to know the Spurling family. It was another highlight for us on our adventure.

A morning photo with fellow campers and workers at Riverview Cabins before we departed.

CHAPTER 18

Day 15: Oden to Fort Smith

"A Nut Is a Nut is a Nut— Unless It's Mr. Peanut."

From Oden, our route turned northwest to Fort Smith, a distance of just under ninety miles. This was a tricky distance, because I didn't know the topography and the distance might require a recharge. Unfortunately, there were few towns along the way. The only possible pit stop for power was the town of Waldron, and it was a bust. If there were a lot of hills or mountains with uphill grades, that could be a problem. It was, in effect, another "hole" in my planning schedule.

I was now more regularly supporting BenHop on his bike by supplying charged batteries to him. Ben used three to four batteries in a day, and he and I never knew when his power would run out. When it did, the bike was so heavy that it was not easy for him to continue pedaling the bike with no power—at least not for long. So I had to be close when he needed the charged batteries so his downtime was minimized.

The benefit for the group was that this freed up the supply truck to be wherever it was needed. Why was this important? Well, it became obvious to us because of the big difference in vehicle-speed capabilities. Because Ben could only do a maximum speed of 25 mph, he could not keep up with the scooters. If the scooters stayed with him, they weren't as efficient. And with Susan trying to continue to plan our upcoming events and do PR, she began telling the group to go ahead of her in the mornings and she would catch up.

This led to BenHop leaving before the rest of the group in the morning. This enabled him to get a fast start on the day, perhaps to have downtime because of a flat tire, or simply to need a rest and take it, yet not to slow the rest of the Tour. If all went right, he might even make it to our destination before everyone else. But this meant that someone had to support him with batteries—either Sean in the supply truck or I.

If the rest or part of the group were not ready when Ben left, their suitcases would likely not be packed, and other equipment could not be loaded in the supply truck, so it would be hard for Sean to support BenHop. So I agreed to support the bike. I recognized that *my* average speed would now be 25 mph also, but there was a side benefit. Power. At the slower speed, the LEAF uses a lot less power, and, on some days like today, that might be crucial. This strategy still required some flexibility, and Sean and I did trade off support for Ben at times, notably whenever I had to drive ahead to recharge at a midday pit stop. Most important, it worked.

We made our way to Waldron and decided to get a bit to eat at the Char Broiler restaurant, which looked like a good hamburger joint. As expected, when I looked around, I saw no place to recharge in the area. So I inquired at the restaurant about a 110v outlet so that I could at least get some power for recharging. While a 110v full charge would take over half a day, whatever I did get into the battery might be enough for me to get to the Nissan dealership in Fort Smith, where I could recharge overnight. I was directed to an outlet under the Char Broiler sign, and that's where I got a needed charge for one and a-half hours.

After lunch, we continued through some rolling hill countryside. I got tired and we pulled off into a small roadside shop called Rock Station Antiques. We didn't think it was open, but when we learned it was, we went inside. Inside was a treasure chest of unique memorabilia. Though we didn't end up buying anything, we had a fascinating talk with the owner, Teresa, who told us the story behind many of the items in her store. Our trivia thirst was quenched.

On we went, and before long, we were pulling into Fort Smith. We'd made pretty good time, and I think we beat the scooters into town. I drove over to Orr Nissan of Fort Smith and had a great conversation with one of the sales managers, Noel, about our trip; he had confirmed to me by phone ahead of time that its charger was operational. Very supportive,

*BenHop finishes swapping out a low-charged battery for
a fresh one that I carried in the LEAF.*

*Mr. Peanut gave us an "assist" by pointing
the way for us to drive into Fort Smith.
"Thanks, Mr. Peanut!"
www.planters.com/mr-peanut*

*Char Broiler sign is
my power source for
lunch.*

Noel offered to help us out in any way he could.

I left the LEAF to charge overnight and hitched a ride with Sean over
to the hotel. Fortunately, I had made the complete ninety-mile leg without a full recharging which, I knew, was due to my slower-than-normal
speed and a favorable topography. I filed this away in my memory bank
for future reference.

CHAPTER 19

Day 16: Fort Smith, Ar. to Okmulgee, Ok.

"An Open-Door Policy Makes Me Sweat!"

We collected the LEAF in the morning from the dealership and set off on a westerly course, which would take us across the Arkansas River and over 100 miles into another new state: Oklahoma. With every new state, our excitement was growing. But the crossing into Oklahoma, like the one into Arkansas, would be a challenge; it involved crossing another highway bridge.

US Route 64 is not a four-lane highway as it crosses the Arkansas River from Fort Smith to West Fort Smith in Oklahoma, but it is a busy one. We gathered as a group once again near the entrance to the bridge. Our strategy was similar to the previous one; I would take the lead, and Sean driving the supply truck would protect the rear to provide maximum safety for the scooters and the cyclist while the documentary crew filmed it all from the front. With less traffic and slower speeds, this crossing was easier—and we were now in Oklahoma!

Riding into Oklahoma had its advantages and disadvantages. On the plus side, the highway was straighter and flatter. On the minus side, there were fewer trees for shade, and the traffic was faster, requiring the riders (and me) to ride on the side of the road much of the time. Whenever we had to do that, there was risk of sharp objects and a flat tire.

We made good time despite the opportunities for delay and eventually

BenHop didn't get the memo (sic) on stopping at the border sign, but the rest of us did.

found our way to a Mexican restaurant, El Jarocho, for lunch. At this point, the group split up because Benswing and I needed to recharge and the closest place to do so was an RV park about thirty miles ahead; we couldn't take the risk of not making it the full distance to Okmulgee, the Tour's stop for the night.

We made it to the KOA campground, and, with a little convincing, the owners agreed that we could use one of the outlets. Ben and I then had the "luxury" of waiting in the air-conditioned office, doing laptop work and talking with the owner. It was times like these when we felt a little guilty since the rest of the group was riding in mid-ninety-degree temperatures outside. And when you're riding unprotected in the hot sun, it's not a lot of fun. Nevertheless, there was not much Benswing and I could do but wait the one and a-half hours on this day to get sufficient charge to complete the day's segment. Some days, that was how it was.

After we were sure we had enough charge to finish our day, Ben and I took off from the RV park and soon made it to the motel for the night. It was late afternoon and thus still hot! The others had not arrived yet, so I decided to continue charging the LEAF. However, there were vehicles in the parking spots in front of my window, and the window didn't open easily, so my only option was to charge the car through my doorway. And, of course, while I was doing that, I couldn't run the air conditioner. So that meant that while the LEAF charged, I sweat.

In fact, more than once, I had to leave my motel door open for the night so the LEAF could charge. One night I even taped the door shut with duct tape so that if anyone tried to enter the room, I'd hear it tear and wake up in time to prevent the perpetrator from stealing my priceless New Balance tennis shoes or, worse yet, using me as an inflatable sex toy.

Paranoid? Maybe. But I was in the Wild, Wild West now.

Our recharging pit stop in Checotah, Oklahoma. Note that we're approaching mid-way in our journey!

One thing that I'm surprised we didn't get grief about was that charging from motel rooms always meant a cord across a walkway. Even if I taped it down, there was the risk of tripping and injury. But no one ever tripped over the cord, and I guess I just got lucky that no one got seriously hurt.

All the others were pretty hot and tired by the time they arrived at the motel. We had dinner and then pretty much packed it in for the night. I continued to charge the LEAF after returning from dinner, and, around midnight, I had a full charge and could finally close my motel door for the night. It was great to have that cool air all to myself!

Left: Charging from my motel room in the hot Oklahoma sun. Top Right: My AeroVironment plug fits perfectly in the A/C outlet. Bottom Right: An open-door policy, and the cord was always a trip risk.

CHAPTER 20

Day 17: Okmulgee to Oklahoma City

"God Bless the Children."

Catfish and camping anyone? Well, I'm not a seafood lover, let alone catfish, but we found this combination during the next leg of our journey. We had another long trip for the day, so Benswing and I needed to recharge at a campground near the small town of Seminole, Oklahoma. We had a project planned in the afternoon, so Ben and I were on our own, and we left early in the morning to recharge. We reached Henryetta around 9:30 a.m. When we arrived, we found the campground okay, but the office was located up front in a restaurant along the highway. And the restaurant, which specialized in catfish meals, wasn't open yet.

We had no idea when the restaurant opened, so Ben and I waited for a while. About 10:00 a.m., we thought we saw some movement inside, so we knocked again on the door. We were greeted by Jamie, the restaurant manager, who wasn't too sure about us until we explained what we were doing, and suddenly she became quite interested in our cause and welcomed us in. She gave us permission to recharge at any campsite that was open, then told us to come back in and she would fix us a catfish meal! Before long, Ben and I were recharging, and we went back inside the restaurant.

Now, I know virtually nothing about catfish, other than they're ugly. And I had never eaten it, or really had any desire to eat catfish. But Julie was being super hospitable, and she had the cook make us up a huge plate of catfish and corn fritters. So how could I turn it down?

Well, it turned out it wasn't bad. Catfish really isn't to my taste buds,

Camping and Catfish!

but it was good, and our conversation with Jamie, who was so nice and interested in our adventures, was well worth the time spent. We could not have been treated more nicely by a stranger than Ms. Jamie. She wouldn't even let us pay for the meal! That was above and beyond, and I hope Jamie reads this book to know how much we enjoyed our visit. If any of you are in the area, stop by, say hello, and check out the albino catfish in the aquarium!

We were making our way to Oklahoma City, about ninety-five miles away, which was our overnight layover. However, we had first planned a stop in Moore, Oklahoma. You may remember that, in the spring of 2013, the city of Moore was hit with a terrible tornado. We all probably saw the images from the media about it; it was a devastating F-5 tornado that tore a huge hole through the city.

We didn't know if there was anything we could do or not, but the tornado had happened only a couple of months before, and we wanted to ask. Sean and Stuart met us at the volunteer center in Moore, and we met with Jason, one of the coordinators.

Jason thanked us for offering to help, gave us a quick tour of the volunteer facilities, and got us registered. He selected a small project for us to help a family that had lost everything in the tornado and that was in the process of moving forward by rebuilding. The lot where the family's house had been, however, still had many pieces of debris that needed to be cleared before new construction could begin, so our task was to clear the debris. We were given gloves, rakes, shovels, a wheelbarrow, and a few other tools, and away we went.

To this point, we had seen the place where the tornado had touched down, but the volunteer center was at the edge of the area hardest hit.

Charging at Round-Up RV Park in Seminole, Oklahoma.

A great catfish-and-fritter lunch with Jamie
and the 750-gallon catfish aquarium.
www.facebook.com/catfish.roundup

Meeting the volunteer coordinator at the center
and checking out the working tool storage area.

Driving over to the address we had been given, we got our first real look at what damage had been done. To hear about it is shocking; to see it in real life is mind-boggling. Following are a few pics of what we saw driving into the hardest hit area:

Houses were in pieces—some completely gone. Others were so badly damaged, with missing roofs, walls, and more, that they were condemned.

Those standing were peppered with shrapnel and had bent garage doors and broken windows. Basically, they had every type of damage you can imagine. Trees were stripped bare of leaves; some areas had ground with no grass or vegetation. In one spot, a hospital sat in shambles with nothing left of it. It was very sad and eerie. I live in Tennessee, an area prone to tornadoes, so the images hit home to think about the awesome power that a tornado can have and the damage it could do to my neighborhood.

We drove down the street to our address and noticed that every home on the street was condemned. The homes were in various forms of damage, but they were all in disarray of one sort or another. At the end of the cul-de-sac, we found the empty lot where the family's house had been. There was nothing there. It had likely been cleared by volunteers or someone else, but there was no structure of the home left. Next door was a condemned house marked with a large X, but we noticed that someone was still living in the house.

We took our tools and started to work. Much of what we had to do was raking to pull paper trash, pieces of concrete or large stones, pieces of wood, and more and moving it all into a center pile for removal. Occasionally, we would find a small memento that we thought might be valuable for the family and would set it aside. I remember finding a baby shoe and a kid's storybook, mostly intact. Tears came to my eyes when I thought how traumatic this had to be for the family. Every bit of the world

Debris from what used to be the local hospital (top) and what's left of a residential area (bottom).

*Neighborhood
of the lot we
drove to with
houses completely
gone and others
condemned.*

they knew had suddenly evaporated in minutes. It was terribly sad. It was painful. It was real.

Stuart, Sean, Benswing, and I worked in the very hot sun for the afternoon, cleaning up the lot. When we were done, we had a small pile of debris that we'd pulled together. Before we went back to the volunteer station, we decided to drive around the area. In just a few blocks, we came across a number of cars and an obvious memorial area. It was the school where seven children had died during the tornado; they had been inside when the tornado flattened the school.

Again, my words cannot do justice to the heartbreaking scene we witnessed that day, which will forever be etched in my memory.

Back at the volunteer center, we met the owner of the property that we had cleaned up. We offered our sympathies for his tragedy while he thanked us for our help. We inquired about how his family had survived, and he related his story to us. They had two kids who had gone off to the school where the seven children had died. As the storm hit, the parents were at home and saw the tornado warnings; the funnel cloud had touched ground and was heading directly for their house! With little time to gather anything, they jumped into their car and drove away from the tornado as fast as they could. They narrowly escaped, but they had no idea how their two children were. Fortunately, while one was trapped for a short while, both of their kids were fine. But the traumatic events the family had experienced were still raw; the owner choked up a couple of times while telling us his story.

I had been in the tornado-ravaged city of Joplin, Missouri, two years earlier, and it had definitely left an impression on me. But that was a few months after the fact, and I had just driven through. This was different. Moore was still fresh: people were still dealing with grief and the impact. People were still living in homes that had been condemned because of damage. The images were haunting. I've never been in a war-torn city (thank God), but it's the only thing I can think of to compare it to. We left

Sean takes a picture of Stuart, Benswing, and me after we complete our small cleanup task.

Moore wishing we had more time to help. Was Moore another element of global warming? Who knows? But if it is, it's scary to think that there will be more Moore, Oklahomas in the future.

This was also the last day we would see Stuart until the end of the trip. He had a job offer from the government of Cambodia to be a consultant on environmental legislation, so he left us to finalize arrangements. It was a good opportunity for him, but I was sad to see him go because he'd become a friend. His sensitive nature had continued to rub some people the wrong way in the group, and I knew Susan was concerned about its affecting morale. But Stuart was smart and had a clear message he was passionate about, yet he hadn't been given many opportunities to contribute as expected. I even had some empathy for Stuart's environmental message now. He was a dedicated environmentalist, and the Tour would miss him.

The heartbreaking scene of seven chairs and favorite toys for the seven schoolchildren killed. https://moore.recovers.org

CHAPTER 21

Day 18: Oklahoma City to Clinton

"A Field of Chairs When Evil Dares."

"Who's Passing Gas?"

The next morning, BenHop and I were the dynamic duo, I helping him with the electric bicycle, while many of the others went to play at a local model-airplane manufacturer that specialized in solar long-distance usage. I decided to let Ben get a head start on me since his batteries usually lasted one–two hours, and I went into downtown Oklahoma City.

Memorial chair for one of the victims.

While it may sound a bit morbid, having just spent the afternoon in Moore, I was curious about another major disaster that happened in the area, the Oklahoma City bombing of the Alfred P. Murrah Federal building on April 19, 1995.

This domestic terrorist bombing destroyed one-third of the building, killed 168, and injured over 680 innocent people. The blast was so powerful it damaged or destroyed 324 buildings, destroyed or burned 86 cars, and caused an estimated $652 million dollars in damage.

The Oklahoma City National Memorial commemorates the victims, survivors, and rescuers involved in the attack and lies on the grounds of the former Murrah building. Annual remembrance services are held at the same day and time as the bombing took place: April 19 at 9:01 a.m.

It was a nice, sunny day as I walked around the memorial. I walked past the large bronze gates bearing the time of the attack and the time recovery started (two minutes apart) and a large reflection pool; then I took in the Field of Empty Chairs.

The Field of Empty chairs is a collection of 168 empty chairs hand-crafted from glass, bronze, and stone to represent those who lost their lives, with a name etched in the glass base of each chair. I learned that the chairs represent the empty chairs at the dinner tables of the victims' families and that they are arranged in nine rows to symbolize the nine floors of the building; each person's chair is on the row (or the floor) on which the person worked or was located when the bomb went off.

The chairs are also grouped according to the blast pattern, with the most chairs nearest the most heavily damaged portion of the building. The nineteen smaller chairs represent the children killed in the bombing. Three unborn children died along with their mothers, and they are listed on their mothers' chairs beneath their mothers' names.

I walked past the Children's Area, where I learned more than 5,000 hand-painted tiles, from all over the United States and Canada, were made by children and sent to Oklahoma City after the bombing. Most are stored in the Memorial's Archives, but a sampling of tiles is on the wall in the Children's Area. Chalkboards provide a place where children can draw and share their feelings.

I wish I'd picked up some chalk that day and expressed my feelings.

Bronze entryway to the Oklahoma City Bombing Memorial notes the times of the attack and the recovery.

The Field of Chairs sits on the site
where the Murrah Building once stood.
https://oklahomacitynationalmemorial.org

Children's Area tiles and
chalkboards.

Having seen two days of memorials, I realized the memorials were a somber reminder of how tragedy can strike at any time. With no one else around me on this morning, I was able to take that thought in and realize how thankful I should be for my life as I know it. Others aren't so fortunate.

I spent about an hour at the memorial and realized I needed to hustle to catch up to BenHop. He was actually waiting for me when I found him along the road ahead, and we swapped out the battery. And off we went again.

It was another hot day in Oklahoma, with the temperature over 100 degrees again. And while I had air conditioning, I rarely used it because I preferred to save power. The A/C used it up—fast. So it was hot for me, too. Not as hot as it was for BenHop and Rachel. As long as they were moving, they were okay. But it was still h-o-t! I just cranked up the stereo a little louder to overcome wind noise.

This leg was another long one, taking us just under 100 miles for the day. And as we hit early afternoon, our route took us past a very old

We made good time despite the heat, passing rivers and miles of grasslands and trees.

gasoline station/hotel called Provine Station. The classic Phillips 66 gasoline pumps were remnants of a filling station built in 1929 and marked our linkup with the famed Will Rogers Highway and Route 66. One of the owners, Lucille Hamon, ran the station for 59 years! A short distance up the road, we found a place to eat—the new Lucille's, which is a step-back-in-time modern diner.

Lucille's is filled with Route 66 and bygone-era memorabilia: from road signs to posters of legends of the 1960s to lots of gasoline-company remnants. The scooters and the motorcycle caught up to us here, so Lucille's was a great place to relax for a couple of hours and enjoy good food.

I used an outside 110v power outlet at Lucille's to charge the LEAF. While it wasn't a lot of power, with the flat ground and slow speeds, I didn't need a lot and was able to avoid an interim charge this time.

This was the beginning of our close tie with the famed Route 66 highway. From here in Weatherford, Oklahoma, to Santa Monica, California, we would either ride on or ride near Route 66. It was a good route for us

BenHop seeks shade at Provine Station, which is listed on the National Register of Historic Places.

because the traffic was low and the sites plentiful—that is, when we could find them. As we would soon learn, sometimes the GPS maps and reality didn't align.

Within a couple of hours, we arrived at our destination of Clinton. I decided to recharge the LEAF at the motel overnight and subsequently cancelled my RV reservation. I had made a few campground reservations along the way in key locations where I had to know I had power available. But by being creative at the motel, I now knew I that I didn't need to be shuttled from the campground to the motel or sleep in my car all night; i.e., I could simply charge at the motel and sleep in a comfortable bed. Well, okay, it was an Econo Lodge, but it beat the LEAF seats!

Lucille's Roadhouse is a chest full of memorabilia from the past—and a good place to eat. http://lucillesroadhouse.com/

Who's passing gas? We are. And in doing so, we're passing the baton. Once a rare gasoline station for automobiles, Lucille's now charged the Nissan LEAF for the next generation of autos.

CHAPTER 22

Day 19: Clinton, Ok. to Shamrock, Tx.

"A Wheelie Important Repair."

"I Missed You, Dog-Gone It!"

Another notable day! The distance was only about eighty-five miles, but we would be heading into a new state—into the panhandle of Texas. It would also be another hot one with the temperature reaching about ninety-seven degrees and still humid.

BenHop got off early in the morning along old Route 66. I left a while later to catch up and eventually found him to support him with batteries for the day. However, Ben was having a problem today—wheel problems. So before long, we hunted up some shade under an overpass, and I called for Sean to bring the supply truck for tools and parts. So much for an easy day.

Sean hustled over with the truck, and soon he and Ben were discussing the next steps. They needed some tools and parts, but getting to them wasn't that easy, because all the luggage, the most utilized part of the cargo, was loaded in the back of the truck. So after pulling out many of the suitcases, they made a path to the front of the truck. Ben was worried about ongoing problems with the wheel, and he and Sean decided to swap Ben's wheel with one of the wheels on the spare bikes we had. Sean, being our volunteer and capable mechanic, was a big help to Ben, and together they made the repairs. It was a time-consuming repair, though, because they had to pull the wheels, tires, and tubes off and rebuild the wheel spokes. Meanwhile, the scooters caught up with us and stayed a while before heading on. All in all, the repairs caused about a two-hour delay in our day. Most important, we were rolling again and moving forward. So far, no problem or delay had been too large to overcome, and we were still on track for our forty-four-day target. Grateful to be moving again, we pressed on.

We made our way into Elk City. It wasn't very large, but it did have a Nissan dealership, and I decided to stop, even though it wasn't listed as having a Level 2 charger on the websites I used. Well, lo and behold, they did have one, so I plugged in. The dealership folks were very nice and suggested we try Lupe's for lunch; they even gave me and Rachel (my rider for the day) a ride to the restaurant, while Benswing and BenHop rode over to it.

Lupe's turned out to be a great lunch spot. Not only was the food fantastic, but we had an extremely friendly waiter, who introduced us to the owners. The owners were a husband-and-wife team that were very interested in our project and quite supportive. In fact, when it came time to pay the bill, they wouldn't take any money, so the least we could do was take a picture in front of their restaurant with them. And I love their "special" sign: 2 Tacos 4 the Price of 2 Tacos, don't you? Trust me, the two tacos are worth the price!

After visiting Lupes, we called the Nissan dealership, and someone came over to pick up Rachel and me. We unplugged the LEAF and were

Smart and talented, Sean became our mechanic for several projects, including the electric bike repairs.

Sean works on the rear wheel of BenHop's bike while Rachel flashes the peace sign.

on our way, but not before taking a picture because the folks at Smith Family Nissan were just as friendly as those at Lupes. It was always great when Nissan dealer employees were friendly to us, even though we were not buying anything from them; that's the way it should be. It's amazing how friendly people always made the day and the trip a shorter one.

As we drove on through town, we came across a couple of intriguing

After a superb lunch, we took a picture with our gracious hosts and our waiter.
www.facebook.com/Lupes-Cocina-and-cantina

Smith Family Nissan treated us like family and
even stopped to take a picture with us.

Route 66 landmarks, including the National Route 66 Museum. We did stop briefly to look around, but time constraints forced us to push our visit to "next time."

On we trucked through the last sections of Oklahoma, through the blistering heat, and past other relics of Route 66. We got a kick out of a huge billboard we saw along the highway—for a free seventy-two-ounce steak in Amarillo, which was our destination for tomorrow. The fine print said that you can get it free if you can eat the entire dinner in one hour!

Rachel and I also stopped at a classic Route 66 bar along the way. However, the Harley-Davidson motorcycle out front and the rusty iron gate on the door gave us pause to enter, so we just took a picture and went on our way.

By now, the scooters were far ahead of us and had already passed into Texas. But the documentary crew was waiting for us at a bar called, appropriately enough, Water Hole #2 in the podunk town of Texola, Oklahoma. Texola is the town on the Oklahoma side of the state line; Benonine is the town on the Texas side. I don't know the population of either one, but I'm sure there are more tumbleweeds than people in both of them.

We had a good cold drink and some yuks with the documentary crew at Water Hole #2. I checked out the souvenirs and found a very cool Route 66 hat. We chatted with the owner about Route 66 history. Then, it was time to move on.

With the state line a short distance away, the documentary crew quickly

A sign size to match the highway's image and one of the last memories from a historic period.

What a deal! This can't be that hard, can it?

"Hey, pretty lady, are you going my way?"

filmed our entry into another milestone state. But after I had stopped to ask them if they were all set, I noticed a pack of mangy-looking dogs gathering along the road across the street. And as I started to accelerate, the dogs suddenly got very excited and started barking and chasing the car. The dogs' behavior was kind of humorous as I reached 25–30 miles per hour and they were still chasing alongside the car. Well, this humor wasn't

BenHop pulls in to Water Hole #2 for a cold drink and brief break.
www.drivingroute66.com/water-hole-2-texola

lost on the documentary crew because I had been miked in the car all day for sound and they heard me teasing the dogs and making comments about how aggressively they were pursuing the car.

Suddenly I got a call from Jonathan. He wanted to film these dogs chasing us. So I stopped the car and backed it way up (there was nothing on this part of the highway at all) to a point near the state line again. Meanwhile, the dogs had regrouped as well and were now watching us from across the road. They looked mean, like they might even have been part coyote, and there must have been ten–twelve of them. And with only a couple of houses around for miles, I figured they were probably a pack of wild dogs.

I asked Jonathan if he was ready to film, and he said he was. I started to accelerate the car, expecting the dogs to follow. This time, though, they just sat there barking. So I stopped, and I began calling them and motioning to them. Rachel was leaning over near me to see if she could call them also. Finally, as I started accelerating, one of the dogs started to chase us, so I accelerated more, the whole time with my head watching the dog and the road. He kept chasing us. Hoping others would follow, I accelerated more and commented to Rachel about "at least we got one" or something like that. But Rachel didn't answer. I glanced over and suddenly realized the passenger door was wide open and Rachel wasn't in the car! With a reactionary "Oh Sh*%!," I slammed on the brakes and looked into the

The documentary crew captures the moments when I suddenly realize Rachel is not in the car with me and I slam on the brakes.

Rachel is left standing defenseless while the wild dogs surround her. Surprisingly, they don't attack but appear to be looking for fun or food! Regardless of their intent, Rachel was temporarily petrified until I backed up to pick her up.

rearview mirror—dreading that the pack of wild dogs would attack Rachel and dreading what I might see.

What I saw made me just burst out laughing uncontrollably. In the rearview mirror, instead of pack of dogs tearing up a defenseless woman, I saw a pack of dogs surrounding Rachel in a circle and looking up to her like they were waiting for their next lesson and a treat! Rachel, of course, was terrified and standing very still, worried about what they might do if she moved. But in my rearview mirror, the scene looked like a teacher holding class in the middle of the road. I laughed so hard the documentary crew later said it was "priceless" listening to me on their end.

Nevertheless, I was worried for her safety and threw the car into reverse. I backed up hurriedly, with the passenger door wide open to pick her up. She got in quickly—as white as a ghost. We laughed about it then, and I apologized. I had no idea she had gotten out of the car to try to attract the dogs' attention. After I made sure she was okay, we tried again, and the dogs did chase us a long distance at surprising speed, so Jonathan got his desired take. And I will laugh about that moment every time I think of it for the rest of my life. So many memories were created on this Tour. Many more to come.

A short while later, we pulled into Shamrock, Texas, for the night, and I charged the LEAF from a motel room again. Shamrock's claim to fame is that it is the "official" St. Patrick's Day celebration location for Texas! In fact, Shamrock also has a town website that describes several renovated Route 66 motels, a 3D movie theatre, and a Pioneer West Museum. Pretty impressive for a town with only 2,000 residents in the middle of the prairie.

But when the first tourist attraction listed on the website is a Tesla Supercharger station, well, let's just say there are no leprechauns or pots of gold in Shamrock, Texas, so we just stayed the one night.

I was lucky to charge in Shamrock. Note: no guests were harmed in the charging of this LEAF.

CHAPTER 23

Day 20: Shamrock to Amarillo

"God Takes Us To New Heights . . . Then Runs Out Of Gas."

"A Big Head, Big Hat, Baby Belly Bust."

The next morning, we headed toward Amarillo, a trip of about 100 miles and, of course, a needed recharging for Benswing and me. Our route was the I-40 service road, which ran alongside the interstate and the current Route 66. A short distance from Shamrock, we took the old Route 66 (current I-40 business route) into the town of McLean. McLean is a small town but with some interesting tourist "attractions," such as

- ➤ The Rattlesnakes sign—a reproduction of a former tall and highly visible exit sign off I-40 that was used by E. Mike Allred back in the 1960s to draw traffic to his gas station and his Regal Reptile Ranch, which held a variety of local critters for Route 66 travelers to enjoy.

- ➤ The Devil's Rope Museum. Where else can you get close up and personal with all the varieties of barbed wire? You can here! And, just for kicks, you'll find some of the old Regal Reptile Ranch rattlesnakes here in jars to view!

- ➤ Painted murals, depicting life around the community along Route 66.

The Sights of McLean.

- Old Route 66 Church of Christ—in existence since 1924.
- Confed Service Station, a restored classic Phillips 66 gasoline station.
- Cactus Inn, one of the remnants of the Route 66 overnight travel stays.

We stopped briefly at a couple of these as we made our way through town—a glimpse back into the glory days of Route 66.

A little further down the road, we came into Alanreed, another small town with an old-time filling station and a "hilltop" (otherwise known as a bump on the prairie) rest area. BenHop and I stopped here for a break from the sun, and I went to a nearby RV park and charged for a while. The store had some interesting Route 66 memorabilia and trinkets, as well as some cute graphics painted on the sides of the building; I think people out on the prairie may have gotten bored, so they took up mural painting as a way to pass along humorous subliminal messages about the frustrations and loves in their life. I guess it worked on us!

During our short stay, we looked around a bit and opened the refrigerator-looking door into the ground. Yes, there was a little underground cooling and storage capability, but, more important, when you're the "high point" on the prairie, having a place to retreat to in the event of a tornado is a smart investment!

A little further down the road, we came to the town of Groom. Here we were welcomed by the "leaning water tower of Britten." Originally a working water tower, it was scheduled for demolition until a local businessman bought it and painted his name on it to advertise his truck stop and tourist information center.

Some of the murals of McLean and the historical Cactus Inn.

We stopped for lunch at The Grill, a small restaurant on old Route 66. The owners welcomed us in for a home-cooked meal and wanted to learn about our travels. And while we dined, we asked if they had a power outlet for me to recharge the LEAF, which led to my plugging into a 110v outlet in the base of The Grill sign out front. That would be all I needed to recharge for the day; I was getting very good mileage from my charges because of my slow speeds (while supporting BenHop on the bike) and the low resistance of the flat prairie lands.

After a delicious lunch, we set out again for Amarillo. On the outskirts

*An early fueling station reminds us gasoline autos were
once new; now we need electric charging stations!*

of town, we passed "the largest cross in the western hemisphere," or so it
was said. We didn't know how tall it was, but we knew it was huge. It turns
out this giant cross is nineteen stories (190 feet) tall and can be seen for
twenty miles! And though we couldn't see them, surrounding the base
of the cross are life-sized statues of the Fourteen Stations of the Cross.
Another tidbit of trivia along Route 66.

These small "Roadside America" trivia sites were important as we made
our way westward. It was no longer about cities and people and famous
landmarks; it was now about anything that stood out from the norm, and
above the flatlands around us. They helped break up the monotony of mile
after mile after mile of open lands. For me, I also had the luxury of satellite
radio, which was a great relief. But for BenHop and the other riders, just
staying focused in the high heat and relatively boring surroundings had to
be a challenge. In fact, I would occasionally ride up to Ben and crank up
the stereo so he could hear the music. If the wind was low, he could hear
it, and it was a break for him. Well, maybe a break for both of us as we put
another mile behind us.

BenHop takes a short break at a Post Office/RV park/ grocery store/rest stop overlooking the prairie.

Jonathan and the crew caught up to us before we got into the city. They filmed a lot of BenHop for a while before they headed on to the Holiday Inn.

Here, the documentary crew films from the front of Ben Hopkins as he pedals across the prairie, with me following. During our trip, the crew would film us from across the highway, from behind, and from numerous angles, which always showed me following Ben. The crew later put these multiple scenes together in a video and set it to the music of "Me and My Arrow" by Harry Nilsson, which was pretty appropriate for a lot of our Western United States trip. In general, wherever Ben went, I went—unless I had to recharge or run an errand or wanted to see something. It worked

Charging the LEAF at an RV park in the Texas panhandle.

BenHop rides into Britten, Texas, where the water tower slants a little to the left in town.

Recharging the LEAF at lunch from an outlet in The Grill restaurant sign. www.thegrill-groomtx.com

BenHop pedals past the nineteen-story giant cross outside of Groom, Texas.

The "Prairie Peace Artist" left his peaceful visions
for all to see along Route 66.

The film crew videos
Ben Hopkins from the
rear of its van.

out fine for the most part—except when Ben got "lost."

But I digress. More "lost" episodes to come.

We were not quite done with our "Roadside America" sights for the day, though. We still had the "prairie peace paintings" and, later, the Jesus Christ Is Lord Travel Center to pass by.

We made Amarillo in late afternoon and headed to the Holiday Inn. We didn't have a planned press event for evening, but we did have an event planned. You may recall that sign earlier about a free seventy-two-ounce steak dinner that was free if you could eat it? Well, Sean had decided he wanted to give it a try, and, for whatever reason, (probably for good TV/video) Jonathan agreed to back him if he failed, which was like a $72 meal cost. We all headed to The Big Texan restaurant to watch the show. And it *was* a show!

The "show" included the personal introduction of all those hardy souls who were taking on the challenge. They were then each given an enormous yellow Big Texan hat to clearly identify them before they were read

This Jesus Christ Is Lord Travel Center truck stop was now abandoned.

For whatever reason, God didn't save this huge Jesus Christ Is Lord Travel Center truck stop; maybe he ran out of gas trying?

the rules for the Big Texan meal.

All participants were seated at a central table, with waiters ready to provide needed support. The participants were ready. Sean was excited and ready to go. We all wished him good luck and sat down to order our own dinners.

Sean then got some final guidance from one of the waiters, and, shortly afterward, with cameras rolling, the bell sounded, and he started in, hoping to down it all in one hour!

As we waited for our meals and ate, we each periodically checked in on Sean to offer helpful encouragement and to see how he was progressing.

At about fifteen minutes in, Sean was doing pretty well. Not eating fast, but eating steady. Steady is good.

At thirty minutes, however, Sean was visibly slowing. The reality that he may have bitten off more than he could chew, so to speak, appeared to be setting in already. And by forty-five minutes, it was basically over. Sean was toast. I believe he opted not to eat more rapidly so as to avoid getting sick. And, in fact, when the hour was up, I think he had only gotten down maybe half of that steak.

He may not have admitted it up front, but he never had a chance at that challenge! Later he told me that he had made a significant error in judgment earlier in the day when some of the Tour members had wanted a bite for lunch and he decided to eat a hamburger or two. To be honest, with the size of that hunk of meat the waiters served, I don't think it mattered. He was a bust. A big, fat baby belly bust! Caught on tape.

While I left to charge the LEAF at the local McGavock Nissan dealership that night, Jonathan charged that steak on his credit card.

And with Sean waddling behind us, we all headed back to the Holiday Inn.

It had been a "full" day.

The full team is here to support Sean in his attempt to conquer the Big Texan meal!

Meal consists of: Shrimp Cocktail, Baked Potato, Salad, with Roll, Butter, and of course the 72 oz. Steak

1. Entire meal must be completed in one hour. If any of the meal is not consumed (swallowed)... YOU LOSE!
2. Before the time starts, you will be allowed to cut into the steak, and take one bite. If the steak tastes good and is cooked to your satisfaction, we will start the time upon your acceptable approval. The time will not stop, and the contest is on, so make SURE before you say "yes."
3. Once you have started you are not allowed to stand up, leave your table, or have anyone else TOUCH the meal.
4. You will be disqualified if anyone assists you in cutting, preparing or eating of your meal. This is YOUR contest.
5. You don't have to eat the fat, but we will judge this.
6. Should you become ill, the contest is over... YOU LOSE! (Please use the container provided as necessary.)
7. You are required to pay the full amount up front; if you win we will refund 100%.
8. You must sit at a table that we assign.
9. If you do not win the steak challenge, you are welcome to take the leftovers with you.
10. No consumption or sharing of the leftovers is allowed in the restaurant once the contest is over.
11. If you fail to complete the challenge, you must pay the full $72 dollars.

Official rules for the "72 oz. Steak" challenge taken from the Big Texan website.

Sean pounds the table— ready for his meal.

Sean gets last-minute guidance from a waiter. Check out the size of his steak!

CHAPTER 24

Day 21: Amarillo, Tx. to Tucumcari, N. M.

"Back to the Future."

"Someone Needs to Ice That Guy."

The next morning, BenHop took off early, as he usually did, while I tried to find someone to take me to the Nissan dealership to pick up the LEAF. Sean was either tied up or still in bed, so I got George to take me. Normally, this was a short ride, but, in this case, the dealership was on the far west side of town, and we had stopped at a Holiday Inn on the east side of town, so it was about a twenty–twenty-five minute drive. The good news was that the dealership lay in the direction we were going and that I was able to locate Ben pretty quickly at the western edge of Amarillo.

Once again, our route would follow the I-40 service road, which was also old Route 66. Just as we left the city limits, we saw the famous Cadillac Ranch on the other side of the freeway from where we were. Unfortunately, there was no easy access from the westbound lanes where we were, so we couldn't stop to check it out. We passed other interesting highlights, like Rooster's Mexican restaurant and the Boot Hill Saloon in Vega, Texas, but we didn't stop. Today's drive was one of our longest, at nearly 115 miles, so we couldn't take the time for stops. The weather was a little better, which was good; it would top out in the mid-90s today—much better than the 108 we saw in Oklahoma!

Distant shot of the famous Cadillac Ranch and
a close up of the big rooster at Rooster's.
https://en.wikipedia.org/wiki/Cadillac_Ranch
www.facebook.com/rooster.arellano

I did make one short stop in Vega, however. I stumbled across an intriguing restored Magnolia gas station. A two-story building made of stone (families often lived above their businesses years ago), the station housed classic memorabilia, and the grounds had been made into a small park for picnics and play.

It seemed ironic that we had now seen several old gasoline stations along our route that once symbolized the frontier of technology with the rapid growth of gasoline-powered automobiles in the United States. Now, here we were with the latest technology in the form of electric vehicles looking for the same type of infrastructure to power our motors, yet it didn't exist yet. Someday it will exist, and, someday, I'm sure, there will be a restored electric charging station which will house similar memorabilia needed to support the cross-country travel needs of Americans who switch to electric-vehicle transportation. Was this a look back to the future?

We slowly pushed onward to the next town. The service road ran directly beside Interstate I-40 in between towns, but, near the town limits, it reverted to the old Route 66 and ran directly through each small town; this gave us the opportunity to see many of the old businesses, such as gas stations, motels, and tourist sites, that either dried up and were vacant or that lost significant sales once the interstate was built and other new businesses sprang up.

One example of a negatively impacted business was the MidPoint Café

Front view and interior of the restored
Magnolia gas station built in 1924.

and the nearby Fabulous 40 Motel, in the town of Adrian. Once thriving establishments, they now struggle to survive. At the café, the group had decided to meet for lunch and discuss preparations for filming as we passed into New Mexico.

It was an appropriate location because the MidPoint Café marks the midpoint of the original historic Route 66 journey from Chicago to Los Angeles; i.e., it is 1,139 miles to each city from that spot. The café would provide a good film opportunity, too, because it's a 1960s-genre eatery. Our meeting there was a good idea, but not all ideas go as planned.

While this activity wasn't a big issue, it indicated one of our Tour's major weaknesses: organization. There were a number of contributing factors, but the end result was a group which basically moved forward despite being pulled in multiple directions. I called this our "amoeba" management approach.

The word *amoeba* comes from an ancient Greek word meaning "to change." The amoeba moves by continually changing its body shape, forming extensions called pseudopods (false feet) into which its body then flows. A bit dysfunctional, it moves on the basis of the direction where pseudopods are formed.

Our group was pulled in multiple directions—from personalities and preferences to differences in our vehicle speeds and unique support needs to our very tight travel schedule and the consumer/press events that were planned and more. At any given time, there was stress because of these pulls on the group. Most of the time, it wasn't a big issue; sometimes the frustrations came to the surface.

Glass-domed antique Magnolia fuel pump.

The Amoeba Management Approach

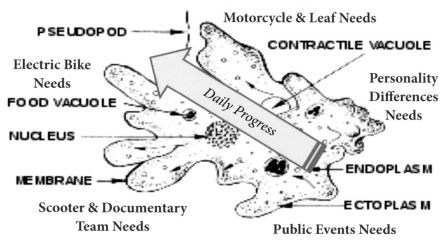

PSEUDOPOD

Motorcycle & Leaf Needs

CONTRACTILE VACUOLE

Electric Bike
Needs

FOOD VACUOLE

Personality
Differences
Needs

Daily Progress

NUCLEUS

ENDOPLASM

MEMBRANE

ECTOPLASM

Scooter & Documentary
Team Needs

Public Events Needs

131

Fabulous 40 Motel and MidPoint Café, which are located
1,139 miles from Chicago and LA.

Susan, as the sole source of the Tour's funds and the original contact for everyone, was the obvious person that everyone looked to for leadership. However, detail was not really one of her strengths, and, with all of the different projects she was involved in (riding her own scooter, press-and-consumer-event-planning-on-the-go, tracking funds, making hotel reservations, and more), she simply had too much on her plate to be in total control. More importantly, Susan never delegated to anyone so that there were clear lines of responsibilities for someone to take charge of each major element of the day's activities.

In retrospect, I could have assisted more. But I had mixed emotions about trying to take control for a project someone else was funding and had planned. Different people had assumed roles as the Tour had progressed, though they had never really been assigned to or asked to assume those roles. I, for example, had become the route coordinator because I had mapped out the entire route with Susan. While, initially, I just told everyone where we would be going for the day, as the group split up because of different vehicle speeds, I made copies every day for all the others so that they could find their way on their own if they needed to. But many activities simply didn't have a clear figurehead.

Our amoeba management style let us down on this day. After BenHop and I rode into Adrian, Texas, we stopped for a drink, and then Ben rode on. However, because I sometimes got bored, I decided to check out the Magnolia Station to see what it was, thinking it might be a train station since I'm a steam-train enthusiast.

It wasn't unusual that I would break away from BenHop occasionally and then catch up with him down the road, since I could travel at much higher speeds in the LEAF. I spent a little time at the station and then headed out to find Ben. In the meantime, I got a message from Jonathan that the group had decided to meet at the MidPoint Café for lunch. That was okay with me, but Ben was already out of town and heading for the New Mexico border. And as BenHop had been prone to do, he had no phone contact; the phone that A2B Bikes had given him was a prepaid phone, which had run out. Ben simply didn't bother to get another, probably because he knew I was always close by. Regardless, I was left having to decide whether to meet with the group for filming or to continue to support BenHop, who would certainly be in need of a recharged battery in short order. And since I didn't want to leave Ben in the heat by himself, I drove on to catch up and support him.

The amoeba style hadn't worked well for everyone; i.e., it continued to move forward for the rest of the group, but it left Ben and me behind. Had one person been in charge of meals and advised everyone the night/morning before departure, everyone would have been on the same page and would have known to stop at the MidPoint Café.

I caught up to Ben, and we pushed on. As we were nearing the New Mexico border, the service road got quirky on us. In one area, it took a big loop away from the interstate and then came back to it. And then the service road simply ended, and we were forced to get onto the interstate. This was a nerve-racking period because even though we rode on the shoulder of the interstate with my emergency flashers on and even though there wasn't much traffic, the traffic that was going westbound was doing 80–90 mph compared to our 20–25 mph. Had a driver not been paying attention and not seen us or drifted off onto the shoulder, the ensuing accident would have been deadly most likely.

Fortunately, we didn't have a long distance before we hit the New Mexico border. Ben and I celebrated by taking each other's picture.

After a short distance further on the interstate, we stopped at the Glen Rio Welcome Center for a pee break. But after our break, we had no choice but to again get on the interstate, and we made our way to the next exit, which was marked Russell's Truck and Travel Center, a large truck and RV stop. This looked like a good place to stop for our lunch while the rest of the Tour was filming at the MidPoint Café.

Ben and I look forward to the "enchantment" of New Mexico. A dinosaur bone from the Glen Rio Welcome Center.

The first point of business at Russell's was to find a power source. So I went inside and hunted up a manager and asked if he had a 220-volt power source. He thought for a bit and then told me I could plug into a place where the ice machines were, around the back of the business. I thanked him for his help and proceeded to pull the LEAF around and back to the loading-dock area and plugged in to recharge.

I went back inside and found Ben. We ordered lunch and checked out the interior decorations. Russell's had a red and silver diner décor complemented with a treasure trove of toy-truck collections and 1950s/1960s memorabilia. There were also murals of celebrities, antiques, and even a free auto museum.

We were just starting our meal when a man walked up to our table and asked if we owned the electric car. I acknowledged I did, and he proceeded to tell me that I needed to move it. I advised him that I had talked to the manager and that he had given me permission to park there so I could charge my LEAF.

He sternly advised me that the manager didn't have the authority to grant me permission to park there and that he didn't want me blocking the dock area. I quickly surmised that this must be the owner, which I respect, but his tone and demeanor were very negative, and I didn't appreciate that. Nevertheless, I went out and unplugged the LEAF and moved the car out

*Charging up the LEAF with the ice machines at
Russell's Truck and Travel Center.*

front. I'm not sure what his issue was because he didn't say—whether I was blocking the dock or he somehow saw an electric car as a threat to his station or he had some other complaint. But the fact that he wouldn't allow me to recharge showed a real lack of any compassion for me as a traveler, contrary to the hospitality he had proclaimed in his advertising.

Ben and I finished lunch and checked out the car museum anyway. It was unique. Small and jam-packed with cars and all kinds of collectibles, the museum was fascinating to walk through. We really enjoyed it—though I wouldn't want the almighty Mr. Russell to know that; yes, indeed, he needed some of that ice in the machines from which I had just unplugged stuffed under his hot little collar.

1960s diner-style café was full of toy trucks and memorabilia.
www.russellsttc.com

Russell's Truck and Travel Center was packed with collector's items.
BenHop enjoys Marilyn's picture.

There were lots and lots of collector items and memorabilia from the 1950s and 1960s. Most of them were in very good condition.

On our way again, we picked up old Route 66, which now ran on the south side of the freeway and began to clearly move away from the new road. It was obvious from the old businesses we passed in this area that the separation from the interstate had dramatically impacted their livelihood; in fact, they simply couldn't survive on Route 66 any longer. Ben and I stopped at some of these abandoned buildings. Old motels, a repair shop, and a fuel station. They stood like ghost towns now—static and empty with nobody taking an interest in them.

Antique cars, drive-up menus, gas station paraphenalia, and a large toy-truck collection at Russell's.

Nothing like a pink Thunderbird to get attention!

After a short while, we headed on down the road. Aside from the abandoned buildings, there wasn't much around to see, though our topography had changed from flat, green prairies in Texas to red clay and sandstone flatlands and mesas.

Old Route 66 came back to I-40 once again, and we rode nearby all the way into Tucumcari, our stopover for the night. We found our motel, and as soon as everyone met up, Sean and I got our suitcases and headed for the room. However, as we went to enter, we spotted a critter that we didn't want to mess with: a yellow scorpion. We didn't want this yellow demon to find us when we weren't expecting it (possibly while we were sleeping!) since I was charging the LEAF from the motel room overnight. So we carefully trapped and released it a distance from the motel.

Ben checks out an abandoned motel on the old Route 66.

Moving into New Mexico brought red clay and distant mesas.

Welcome to Tucumcari!

CHAPTER 25

Day 22: Tucumcari to Las Vegas (N. M.)

"From Lost in Translation to Lost in Transmission."

"Running on Empty. . . Running On. . . .Running Into the Sun, But I'm Running Behind."

www.youtube.com/watch?v=zdHg4QEmBvk

As we awoke to another hot, sunny day in Tucumcari, I think we all realized that we were heading into one of the most dangerous stretches of our Tour. From New Mexico to Arizona and then on into eastern California, we would travel mostly through desert areas with few towns and little population—and thus limited opportunities to find water and food along the route, as well. We had discussed the need for getting out early in the day and, in general, the pairings of riders that had worked out pretty well as we followed along the service road of I-40. At least, until the service road ran us onto the interstate or made a turn we didn't catch because the turn wasn't well marked.

This day, however, we were traveling off the I-40 path and heading north to Las Vegas, New Mexico, en route to Santa Fe the next day. Susan had found some eager communities that wanted to do events with us for electric and solar power, so we would deviate from I-40 through most of New Mexico before picking it back up in Albuquerque in the western part of the state. To do this left us with some challenges, however. Our trip would be another long one again, at over 110 miles. And for me (and Benswing), it meant a recharge. I could find only one place to possibly get 220v power, and the info about that source was sketchy at best. It was, in effect, another "hole" in my power schedule.

That one source was located at a state park campground at Conchas Lake, New Mexico. All I could find online suggested that a campground was there, but where exactly it was along one of New Mexico's largest lakes was unclear. There were eleven miles of park along the Conchas River and fourteen miles of park along the South Canadian River where the campground might be. You might be thinking, just put the name of the campground into the GPS, and the GPS will find it. Well, not so fast, Tonto. As you might find out if you travel through some of the desert states, the Lone Ranger hasn't mapped out all of the roads and the locations with very few people residing there. And trust me, New Mexico is one of those states.

I had two major concerns about this leg of the trip:

1) Finding the campground—that is, where it would be located inside the state park; since the state park wasn't along the line with our route to Las Vegas, how much power it would take to get to/from the park to our route.

2) While the first two-thirds of the leg was relatively flat, a 2,200 foot elevation gain lay in the last one-third of the leg, from the small town of Trementina up to Las Vegas (and who knew about individual peak heights in between?).

With a long distance to travel and a big elevation gain, I knew a recharge was critical—there was no choice. And if I couldn't find this remote campground easily or if there was no 220-volt power source, I would have a big problem to deal with. Running out of power while climbing a mountain was my biggest fear—and my biggest challenge to achieving my goal of crossing the country without using the backup generator for power.

Because of our concerns, it was decided that Benswing and I would go ahead of the group, find the campground and recharge, and then meet up with everyone else along the way. Sean would support the A2B electric

Restored 1960s businesses are numerous in Tucumcari.

More Route 66 businesses looking for tourists in Tucumcari, New Mexico. www.visittucumcarinm.com

bike (with BenHop riding it) and the scooters, both of which would need batteries along the way. We had a plan.

Rachel would be my rider on this day as Benswing and I started out. In the first early-morning hour, we were still on Route 66, passing more businesses that commemorated the yesteryears of Route 66 in downtown

Tucumcari. There were a number of them—restored to the way they were in the 1960s, but still operating today.

This was, in fact, what we had expected to see along Route 66—basically a celebration of the highway's history and success. But those celebration sites and tourist traps were actually very sporadic, almost as if they were totally dependent upon the local inhabitants' support for the tourist trade, or not. And I guess, given peoples' time constraints and the growth of the airline industry, perhaps we expected too much.

We soon turned off Route 66 onto New Mexico Highway 104, a two-lane highway which roughly parallels the Canadian River basin north-south through the state. There was nothing exciting about this road. I mean, *nothing*. No traffic. Nothing but desert rock and distant mesas to see. And I swear, the straightest road stretch in the United States. I would bet money that I could have put the steering wheel between my legs, gone to sleep for an hour, and woken up still in the right lane of this road. And as I think about it, I should have tried it.

I was happy to have Rachel on board and my satellite radio to fight the boredom challenge of this road. Even our cell phones didn't help. In fact, they didn't work at all. Period. Verizon had served me well up to this point, but not in New Mexico. And I soon learned why. The state of New Mexico has private telephone companies still remaining, so cell phone service providers have numerous "holes" or "dead zones" in the state. The only way to call anyone is to use the local phone-company lines. This presented a new twist to the day that we didn't expect. We had no way to communicate with one another—not good since we'd become totally

Only desert and mesas—along the straightest road in the United States?

It wasn't pretty, but this campsite had power, and that's all we cared about.

dependent on electronics for all communications and navigation. With no phones, we were now on our own.

Benswing and I saw the turnoff to Conchas Lake and headed into the park. Since it was a weekday, there were hardly any people around, including park personnel. So we just kept heading deeper into the park and found the dam. The road passes over the dam, which was interesting and scenic, but we still didn't see signs for the campground. We continued on the same road as it wound around the lake for what seemed like an hour. Finally, I saw some trailers, and we headed toward them and found the campground—well, sort of.

There were a number of trailers in the campground, but we saw no one. The trailers that were there looked like they had been there for some time, and some were not in very good condition. The campground itself left a lot to be desired. The sites hadn't been mowed in very long time. Much of the equipment looked like it hadn't been used in years. Rainstorms had left visible ruts through the site parking and camping areas. It almost looked like the place was abandoned. Nevertheless, we decided to find a campsite and try the outlets at the junction box and hoped they still carried power that we could use to recharge.

Looking around, I found one I could pull the LEAF into and get close enough for my charge cord to reach the junction box. After checking our connections we plugged in, and "whaddya know!" it still worked! Now all we had to do was sit around for two–three hours collecting desert rocks and avoiding scorpions. Lovely . . .

Those couple of hours were, well, boring, to say the least. With no power. No cell phone access. No Wi-Fi. No air conditioning. No place for shade or to sit other than inside the car. No people around. So we just walked around the campground, talked, and had a drink or two. We

Our deluxe accommodations included a no-shade tree and pass-through water drainage!

couldn't wait for the charge indicator to read "fully charged"!

Finally, we were charged up and ready to go again. We retraced our drive back to the main road and headed back toward the dam. About this time, we saw the support truck driving toward us in a hurry; Sean was very glad to see us. I could tell something was wrong, though, because he looked anxious. He had the same phone problem we had and told us he had been supporting BenHop on the bike. Susan and Dominique, however, had left via a different route and at a different time, and Sean had assumed they were ahead of him—likely needing batteries. He had left BenHop behind to look for Susan and Dominique and had come all the way to the lake and the campground but had never seen them.

Because the sun was hot and there were too few places for Susan and Dominique to really go off the road for Sean to have missed them, he was nearly panicked that they might be by themselves—perhaps their scooters broken down or out of power. And there was no way to tell anyone where they were. I shared his concern, quite frankly. So Benswing and I found a nearby office, and, luckily, two park rangers had just driven up. We told them what we were doing and asked if we could use the local phone to call Susan and Dominique. They let us use the phone, but we didn't get an answer from either one. This was strange—and even more concerning. Because we had only a few options, I told Sean to backtrack south all the way to Tucumcari if he had to and to make sure the girls weren't anywhere on that part of the road. Meanwhile, Benswing and I would continue north toward our destination and locate them if they were stranded somewhere ahead of us. I was worried that if we didn't find them along the route we'd chosen, it would not be good. It likely meant that something

*If we washed our clothes in the public restroom sinks,
we had clothesline! Or was it telegraph line?*

bad had happened and that we had no way to know.

We set out again in the direction of the dam and were about to cross it when I thought I saw something off to my right—something shiny, something silvery, like a scooter!

Out of nowhere, there they were, parked near some small trees and shade. Both Dominique and Susan didn't notice us pull up because they had laid down to rest. How Sean missed them or where they had been didn't matter. I was just glad that we had found them and they were safe. We woke them up, and they were equally glad to see us, too. They didn't know how to find us when they realized they had no phone service. This was a potential disaster averted.

After a few minutes of comparing notes, we rode together across the dam and back to New Mexico Route 104. We headed north toward Las Vegas and soon got a second stroke of luck: we spotted BenHop pedaling along ahead of us. We were back together again as a group! And we planned to stay that way—at least for the rest of this leg, anyway.

We continued along Route 104 through more desert flatland, with a few hills now and then. For the most part, the only traffic we saw was an occaisional tumbleweed. It was pretty easy, and it was good to be riding together as a group again. With no major repairs or downtime, we simply just kept putting mile after mile behind us, and swapping out batteries.

As we reached the small town of Trementina, we were finally approaching those mesas that we had seen in the far distance a few hours back. Now we would start to climb for that last one-third leg to Las Vegas. And start to climb we did. Not just a gradual incline, but a fairly steep grade up the mesa. A difficult climb for the scooters even and particularly for BenHop

We found the girls resting in the shade; that's Susan curled up to the right of her scooter.

The girls head back across the dam toward the main road to Las Vegas, New Mexico.

on the bike. I had plenty of power to climb the grade, but it came at a steep cost—battery life. And given that I had already had to recharge two-thirds of the distance from Las Vegas, I began to worry about how much battery life it would take to get to the top of the mesa.

We continued climbing, relentlessly winding our way around the side of the mesa. And every mile we climbed, my battery life was falling. And falling. This went on for miles. Finally, I saw a sign for a small town called Alta Vista, or "High View." "That's good," I thought. However, we were still climbing, and I had just gotten a new dashboard message from the LEAF that I didn't want to see: **"Battery Level is Low."**

This sent me into a brief panic period, where I tried to figure out what I could do differently to save energy. The question now became whether I had enough battery life to make it to the top; it was going to be close. I slowed the car even more to save power. I was hoping that the car didn't

Susan and Dominique cross the dam at Conchas Lake.
www.emnrd.state.nm.us/spd/conchaslakestatepark.html

"Traffic" on New Mexico Route 104 (above), and
Dominique relaxes with music on the scooter (below).

BenHop powers up the steep grade toward Las Vegas.

go into "turtle mode," where it essentially shuts down and only utilizes a minimal amount of power to enable the driver to get to a safe place to recharge. If it went into "turtle mode" while we were still climbing, I probably wouldn't make it to the top.

I thought, "I think I can, I think I can make it to the top." I was squirming in my seat beside Rachel, looking for the top of the dang mesa. We were both worried that we wouldn't make it. We were, as the song goes, "running on empty." Finally, we came up to a big, long, sweeping curve, and I realized we were at the top. We had made it! We reached the peak, and the road leveled out. "Whew! That was close!" I thought and breathed out a huge sigh of relief. My worries weren't totally over, however, because there were still several miles to Las Vegas. Fortunately, there were downhill stretches before the land leveled out, so I took full advantage of the car's regenerative braking (recharging of the battery when the car goes downhill and is "pushed" by gravity); this ultimately enabled me to get all the way to Las Vegas with a low-battery power level. It was like regaining the power I had lost going uphill, and I needed every bit of it.

Just after we passed the peak in Alta Vista, the group pulled off the road into a small parking area in front of a home. We took time to relax, grab a bite to eat, and drink together. It was like an afternoon picnic—much needed after a long, stressful day. Everybody was in good spirits, though, and we swapped stories.

As we left our picnic area, the scooters and Benswing went on ahead because we had a planned consumer event that evening in Las Vegas. I stayed with BenHop for support, and we pedaled/motored along the remaining twenty-five miles into Vegas. While the ride was easy, we soon found ourselves trying to outrun a rainstorm headed our way. We narrowly missed getting soaked—well, BenHop anyway.

Ben and I went directly to the motel and had just enough time to check in and clean up a bit before we had to leave for the event, which was being

Getting close to the top of the mesa while LEAF battery life was dwindling.

BenHop tries to outrun a sudden rainstorm as we enter Las Vegas, N.M.

BenHop explains some of the features of the A2B bike with interested consumers.

held in a park close to the motel. Here we met a small crowd that came out to see us and talk about our trip and purpose, and a few took test drives on Ben's A2B electric bike.

For entertainment, Dominique sang songs while we talked with folks about the benefits of moving to electric vehicles. Most of the attendees were people who had been considering electric, so the conversations were usually questions about performance of the vehicles and what concerns the people had.

Good talks.

As it turned dark, the crowd began to disperse, and we headed to the motel. Once again, the turnout was small but the people we did meet were very enthusiastic about our cause. This positive reinforcement kept us upbeat in the daily grind.

Dominique sings her heart out, while local electric-vehicle enthusiasts listen.

CHAPTER 26

Day: 23: Las Vegas to Santa Fe

"Coming Up Flat in Santa Fe." "Solar Power Leads the Way." "Stuck With Stucco Bills to Pay."

The next leg of our journey was a very comfortable seventy miles, give or take. Even for BenHop on the A2B bike, this should be only about a four-hour ride. These days were nice because we had time to stop periodically and enjoy the scenery a bit—"smell the flowers," so to speak. We did have an evening event planned in Santa Fe, our destination, but we still had a more relaxing pace to our ride—that meant a little less stress.

I had never been to Las Vegas, New Mexico, before, so as I started this day, I drove around the downtown area just to see what was there. Though there was nothing spectacular, there were some older homes with great character to them, which I always found interesting. You never know what you might find when you take side excursions through a downtown; sometimes there's some great history or a unique event or memorable people that you find along the way.

After cruising downtown on this morning, I headed out to find BenHop and Rachel—both riding electric bikes today. And since we were following the only road in the area along the interstate, finding them wasn't too difficult. They'd gotten a lead on me by leaving earlier, but I easily caught

Homes of "character" in downtown Las Vegas, New Mexico.

up. And before long, they both needed one of the batteries I was carrying for them.

Rachel had started on the scooters as we left Charleston, but she was now rotating day to day; i.e., one day she rode with me; another day, she rode in the supply truck; now she was enjoying the extra electric bike that we had. The combination of being a people person and variety was right up her alley.

After swapping out the batteries, we continued our journey. Being at the higher elevations of the plateaus had its advantages; this was one of the nicer days we had had the whole trip, with temperatures between seventy and eighty-four degrees during the day. This made the ride a pleasant and scenic one, even if there was desert all around us. And with traffic pretty minimal, it was an easy ride, as well.

Most of the ride was rolling hills and relatively flat overall, although Santa Fe is about 750 feet higher in elevation than Las Vegas at 7,200 feet. So, with good weather, flat terrain, and a good road, this leg was going fast.

As we neared the Santa Fe area, we saw a lot of other bicyclists. In fact, we stopped at a bike store along one of the trails because BenHop wanted to pick up a few spare parts for his bike. The store was seemingly in a remote area, but there were lots of customers! So it must have been a big off-road and on-road bike area. I trolled the store while Ben found his parts. I spied some interesting items, but nothing the LEAF needed.

It wasn't too long after we left the bicycle shop that BenHop got a flat tire. This wasn't unusual. And Ben fairly quickly had the tire off, repaired the hole, and reassembled the bike so we could be on our way again. We rode on into town and began navigating the streets busy with traffic. But as we headed down one of the hills in Santa Fe, I saw Ben pulling off to the side. He had a second flat tire!

Rachel pauses after changing out her battery.

Following BenHop and Rachel along a road paralleling I-25 in north central New Mexico.

With a new recharged battery, Rachel is off again.

One of the interesting bicycle-accessory displays at the Santa Fe bike shop we stopped at.

We pulled over into a shopping plaza parking lot, and Ben began tearing down the bike again. The second flat tire meant that Ben had to look closer to see if whatever had given him the first flat was still in the tire and had caused the second. We called Sean to bring over the supply truck for parts to replace the tube, and possibly the tire, too. This took some time and put us behind the rest of the group. Ben was frustrated a bit, but there's not much you can do. There are hazards everywhere along the road that can cause a flat.

After about an hour's delay, we were moving again and drove over to the Santa Fe Railyard, a hip new area for social gatherings at a train station beside the Second Street Brewery. Susan had arranged another consumer event with the local electric-vehicle and solar-power enthusiasts. After we arrived, we parked our vehicles for display.

We were all set for an evening to talk about electric vehicles and our adventures when our nemesis returned—another rainstorm washed out our event. We tried to park in a covered waiting area, but few people stuck around for the event. It also got cool, so we huddled in the Second Street Brewery and had dinner with the local solar-power enthusiasts. It was fun, but I was cold and wet and wishing I wasn't there.

After dinner, I needed to set up for a recharge of the LEAF. One of the solar-power enthusiasts was more than happy to help. She knew of a

Too bad we didn't have time for a Bobcat Bite!

The calm before the storm: talking with electric- and solar-power enthusiasts before the storm sends us scattering. Rachel and Susan talk to organizers, while Evan and George try to get some video before the rain.

solar-powered level 2 charger near our Sage Hotel, so I drove over there with her. It was pouring rain, but fortunately the charger had a solar cover which protected me from most of it. Still, plugging in made me nervous, not only from the rain but from knowing that the charger was solar powered and hoping it would fully charge the car overnight as I needed. But once the LEAF was plugged in, the gauge said it was charging, and I left it there overnight. In the morning, the car was, indeed, fully charged. Solar power brightened my day, after a very cold and wet evening!

As we were returning to the hotel, I noticed a little commotion along the entryway just to the left of the hotel registration/lobby area. People were outside on the second-floor balcony, and others were standing around below it.

The hotel had a bit of an odd entryway, with the lobby on the right

Rained out for an evening discussion on alternative power sources,
we headed for the Second Street Brewery.
www.railyardsantafe.com/

side and hotel rooms built two stories high running beyond the lobby, more double-decked rooms across the back and also across the entryway to the left side of the lobby. But, with two levels, the upper level on the left protruded over the entryway. It was odd because you had to turn left as you pulled into the center area of the hotel and close to the hotel rooms on the left side.

A few people were sweeping up some stucco on the driveway on this left side, and there was obvious damage to the hotel's second floor. My immediate thought was there must have been an earthquake. Well, there was an earthquake of sorts for those upper-story hotel visitors. It wasn't a ground-shaking quake; rather it was a Sean-quake!

I soon learned that as Sean was bringing in the supply truck, he was either not paying attention or had misjudged the balcony area, and the truck had struck the corner of the second-story balcony. In doing so, he not only gave a big jolt to the residents; he also tore a big hole in the front left corner of the U-Haul!

I roomed with Sean that night, and I gave him some grief in a humorous way. I didn't overdo it, though, because he felt bad enough about it. In fact, he spent a couple of hours the next morning duct-taping the hole in the U-Haul. So we had an asterisked-shaped patch to the U-Haul roof for the remainder of our trip. Thank goodness Susan had purchased insurance on the truck!

But even with insurance, I still don't know who got the hotel repair bill—i.e., stuck with the stucco!

CHAPTER 27

Day 24: Santa Fe to Albuquerque

"Wild Hog Laughs and Coal Mine Shafts."

The artsy air we saw in Santa Fe was difficult to leave; it was fresh and new, reflecting the blending of American Indian culture with modern technology. But we had selected another intriguing route from Santa Fe to Albuquerque, which would be almost the opposite, one rustic and full of history, i.e., the Turquoise Trail.

A U.S. Scenic Highway running the seventy-mile length of our leg for this day, the Turquoise Trail is synonymous with American Indian spirituality, Spanish explorers, adventurous mining, and brave pioneers. A major Western gold rush occurred here around 1825, years before the California Gold Rush. But the Cerrillos, (the little hills) were coveted long before the search for gold for its rich deposits of turquoise, as well as lead ores used to glaze and decorate traditional Rio Grande pottery. The Cerrillos are known for historic stone maul and pick-and-shovel mining in the Southwest.

I traveled with the scooters most of the day on a very leisurely ride past numerous outdoor and indoor studios of artists—many of them seemingly "out there," both in terms of geography and personal preference. These folks must really want their own frontier to settle in: most of the terrain is rocky desert, and they have very few neighbors.

The first small town we came to was the old mining town, Cerrillos. An

Heading to the Turquoise Trail, the riders watched for traffic, and open-range cattle, too!

early source for rare turquoise, which was readily traded in the Southwest, it was rediscovered in 1879, when deposits of lead were found. What started out as a tent city soon became a boom town of hotels, saloons, dance halls, shops, homes, a church, a school, and other supply stores. However, almost as quickly as Cerrillos grew, its mines began to run out of lead around 1900, and the town rapidly disappeared. Today it has only about 200 residents. Only a few shops and galleries remain, along with the Cerrillos Turquoise Mining Museum with its hundreds of artifacts from the Old West and the Cerrillos Mining District.

We spent a little time looking around in an interesting gem-and-turquoise shop and started to explore a couple of streets off Main Street—that is, until we realized that those streets soon became rain-washed, rutted dirt alleys that might lead to a significant delay in our plans for the day. We moved on.

As we continued south, we would periodically spot abandoned old mines along the hillsides above us. Unfortunately, we didn't have time to

The Turquoise Trail is lined with artist studios, including this sculpture garden (far right).

explore them, but they were intriguing. There was a lot of history hidden in those old mine shafts and stories that will never get told, I suppose.

Our next town was Madrid. No, I was informed, not Muh-drid, as in the capital of Spain, but M-a-a-d-rid. So M-a-a-d-rid is a very old coal mining town that dried up back in the 1950s and 1960s with the growth

*Our first mining town along the Turquoise Trail
was Cerrillos, nearly a ghost town now.*

*The owner of the Casa Grande Trading Post and
Mining Museum shares our moment at her store.*

*The Turquoise Mining Museum has hundreds of artifacts from over a
hundred years of mining in the Cerrillos.*

As we explored a bit, we found this homeowner's front gate.
Knock before you enter!

Sean follows the bikers as we pass the Madrid Coal Mines.

of diesel locomotives.

Today, however, it is a haven for artists of all types—from painters to sculptors and glassworkers with fascinating creations in jewelry, furniture, and even clothing. Everything in between, as well, including the odd and the unique. The little shops we found in M-a-a-d-rid seemed to have all these creations, and we spent time simply checking them out.

Lunchtime came, and we bellied up to the Mine Shaft Tavern and ordered meals. The tavern is located at the entrance to an old mine shaft, which is called the Ghost Town Museum and which contains various memorabilia and photos. Adjacent to the museum is the Engine House Theatre Museum, which used to house the steam engines for coal hauling; in fact, you can step into Engine 769, which is part of the theatre itself. The theatre has hosted numerous plays and musicals over the years and houses clips from various movies that have been filmed in M-a-a-d-rid, such as a scene in the movie *Wild Hogs*.

We had a tasty meal at the tavern, probably the best eatery in town,

Lunch took place at the Mine Shaft Tavern and the Ghost Town Museum. Also pictured: the Engine House Theatre Museum.
www.themineshafttavern.com   www.turquoisetrail.org
www.visitmadridnm.com

and listened to music. Usually there are local artists who play and sing during peak hours at the tavern; it's a mix of a locals hangout and a tourist attraction.

Most of the Tour had a beer or two and relaxed while a few of us did more window shopping. Fortunately for us, there wasn't too much drinking, and no fights broke out like the one in the movie *Wild Hogs*!

If you're not familiar with it, *Wild Hogs* is a comedy about four friends (John Travolta, Tim Allen, William H. Macy, and Martin Lawrence) going through midlife crises who take off on Harley-Davidson motorcycles across the country. Among their many humorous adventures, they get into a fight with a biker gang at Maggie's Diner—which is the building shown below in M-a-a-d-rid. It's not actually a diner; it's a clothing store. But the owners keep the diner image up because it draws lots of customers!

We continued down the Turquoise Trail past Placer Mountain and the back side of the scenic Sandia Peak. This part of the trail was mostly downhill—all the way into Albuquerque—so the ride was easy and fast. The area is very pretty near Sandia Peak, but, at the time, we were not aware of the tram ride on the front side of the mountain, which takes passengers over some beautiful rocky canyons. http://sandiapeak.com

Susan had set us up for another consumer event, which was on the west side of Albuquerque. We drove into the city in late afternoon, just in time for early rush hour traffic, and crossed from east to west via a road that took forever to get to the other side of the city. But we eventually made it to the other side and pulled up to the lone electric-vehicle charging station in the area. Here we met with some of the local electric-vehicle and solar-power enthusiasts whose promotional efforts generated a small crowd that we let ride the electric bikes and with whom we talked about the LEAF.

Maggie's Diner from the movie Wild Hogs.
https://en.wikipedia.org/wiki/Wild_Hogs

Ever been to
a Photo Park?
M-a-a-d-rid has
one!

Scenes from a couple of the local shops in M-a-a-d-rid.
Artists on display!

Not a big turnout, but every enthusiast counted.

After about an hour and a half, we headed on to the hotel to call it a day. Truth be told, it was another uneventful day, and the ease of it was not lost on us as we headed off to bed.

Visitors check out the LEAF and talk about how electric vehicles can benefit them.

BenHop and Rachel swap out a battery so intrigued visitors can take a ride.

Sean decided that the best use of his time at the event was to catch a few Z-Z-Z-Z-s.

CHAPTER 28

Day 25: Albuquerque to Grants

"A LEAF Doesn't Always Need Water to Stay Green."

Coming out of Albuquerque, you would think that we'd be headed down in elevation, since Albuquerque's elevation is around 5,300 feet. But, actually, our trip today of some eighty-five miles would take us up again to 6,500 feet across miles of flat, open, rocky desert. Just for good measure, I went over early to the public recharging area and "topped off" the battery before leaving the city.

Most of today, we would parallel Interstate 40, which was generally straight and had little traffic. Rachel was again joining BenHop on the electric bike, and I was support. Rachel seemed to enjoy rotating between the bike, the scooters, my LEAF, and the supply truck with Sean driving it. But today was a bike day, and she had her customary backpack with supplies and extra clothing for the day.

Among her items, Rachel carried a Luci light, an interesting solar-powered lamp that we used the first night of camping and that had come in very handy. The Luci light could be charged up during the day with sunlight and used at night like a gasoline-powered lantern. So Rachel usually rode with her Luci light in tow. https://mpowerd.com

To top it off, it was a very warm day (mid-eighties), and the humidity kept rising into the low nineties. So, for the riders, it was draining: heat, humidity, and few areas of shade. Occasionally, the old highway would break away from the interstate just enough to go through some rock

Paralleling I-40 on old Route 66 for miles and miles.

Rachel rides with her plastic Luci light (far right) and her sheriff's badge from Big Texan Restaurant.

outcroppings where we could find something interesting. But, by midafternoon, the heat was wearing on BenHop and Rachel.

The scenery for the day wasn't all that exciting—a few trailers or run-down houses here or there and maybe a mine or two, but mostly just desert.

With the humidity so high, it was just a matter of time before a quick thunderstorm came up. The rain and even a little hail sent Ben and Rachel scurrying to get their windbreakers on. Unfortunately for them, there was no place to go in the area unless they jumped into the LEAF with me, and had the thunderstorm been bad enough, they certainly would have. Instead, they chose to ride on.

Despite the rain, we keep moving.

On this day, a long train in the distance was part of the scenic beauty we got to enjoy.

BenHop and Rachel stop in a little shade, but later I had to revive Ben with a good towel wave!

The rain didn't last too long, but, in the desert, a rain can significantly cool the temperature. The areas ahead of us didn't have rain, though, so the temperature stayed fairly warm, and the riders didn't have to put on heavier clothing.

On we went through lots of rock and desert, sprinkled with an occasional store or small town. Most of the land consisted of large cattle ranches, but, honestly, there wasn't much here. In fact, a lot of areas here were American Indian reservations, which, sadly, I'm sure, were a far cry from the lands the Indians used to live on.

Thousands of years ago, volcanos covered the area, but only the bases of them (the mesas) are left today because of wind and water erosion. Unfortunately, not much grows in this type of rock and soil.

The long and level open range allows locomotives to pull a large number of railroad cars across the state.

As we approached our destination in Grants, we were thankful for another day without an incident, a day with little traffic, and, aside from a little rain, a pretty calm day.

BenHop and Rachel brave the rain and small hail (above), while (below) Rachel is wondering why I'm not getting out of the car. Answer: A LEAF doesn't always need water to stay green!

There were no public chargers in this area of the state. None. And no RV parks close by, so that meant another motel-charging night for the LEAF. Tomorrow, we would break another milestone as we crossed into Arizona. These uneventful days bunched together would keep us pushing on. Arizona, here we come!

During one stop, Rachel wanted to take my picture, so "Welcome to New Mexico, everyone!"

Rachel stops for a battery change in front of igneous and columnar basalt volcanic rock formations.

Even in the small towns in the desert, there were still signs of people and their faith in God.

CHAPTER 29

Day 26: Grants, N. M. to Chambers, Az.

"Road Maze Days."

"Shootout at the ~~OK Corral~~ Arizona State Line"

Paybacks are a bitch, right? Or so they say. But they are not always immediate.

Way back in either Oklahoma or Arkansas, where it was really hot and we were on the road, there had been a skirmish. A brief release of tension by some of the Tour group who had weapons of mass . . . consumption; i.e., water bottles. Yes, the perpetrators had sent everyone scurrying, including me, for cover as they lashed out with their dirty, germ-filled, and unfiltered sprays. I was basically defenseless at the time, so I just ducked into the LEAF. Sean, often my roommate, wasn't quite so lucky and took several, hard body shots. They were a major blow to his ego. And we never forgot.

But when you retaliate, it's important to A) invoke the element of surprise and B) make sure you have a bigger weapon than your foe. The water altercation had not been forgotten, and when we were in Amarillo, I sent Sean on a mission: to find us a pair of bigger weapons. He did, and he subsequently hid them in the supply truck until we were ready to use them. That wasn't easy, because anyone and everyone was inside the supply truck at any given time.

Today looked like a good day for a battle. It would be hot—in the eighties or so. And we would all be together for the crossing into Arizona. I told Jonathan, our documentary director, to be prepared that, at some point during the day, we would stop the group and that he should be ready to film because "something" was going to happen. He didn't know what, but, of course, he went along with it. Anything for good film, right? The stage was set. Sean and I hid our weapons in places for easy access. And waited. The right time would come.

Our ride today would be another long one of over 105 miles, requiring Benswing and me to recharge along the way. We had identified an RV park in Gallup, New Mexico, and that was our initial target. Shortly after we left the motel, however, I got a call from a member of the New Mexico Solar Energy Association in Gallup. This fellow, let's call him Jack, had apparently heard about us from the Santa Fe group and wanted to know the technical specifications and the connection I had for my LEAF because he wanted to charge the car when I was passing through. That sounded good enough, except that it wasn't my car. So I told him he would have to contact Nissan to get its approval because I didn't know what amperage or voltage he would be charging at and there had been cases where LEAFs that had used third-party chargers had actually damaged their batteries—that is no small cost. I knew that a new battery would probably cost $10,000. In addition, a badly damaged battery would all but stop my Tour. So Jack went away grumbling, but it wasn't the last I'd hear from him.

Meanwhile, I contacted my bud, John Arnesen, back at Nissan to get his input. He wasn't so thrilled about Jack's idea and told me it was a risk. Probably not a big one, but a risk. If I decided to let Jack charge the LEAF, it would be my risk.

By this time, we had made our way on Route 66 to just north of the small town of Thoreau and to a local store/souvenir shop located at the Continental Divide.

We spent a little time checking out the trading stores with a lot of American Indian jewelry and handmade items. There was also some interesting history here, including trading companies, a Top o' the World Hotel, local farms which were irrigated for carrots in the 1940s, and sporadic uranium mines. All interesting, but ultimately it was time to move on.

Having climbed some 800 feet from Grants to the Continental Divide, we all now knew it was now time to go back down 800 feet into Gallup.

The early group takes time out for a picture at the Continental Divide.

*There are small trading stores and some interesting history at
the Continental Divide.*
www.legendsofamerica.com/nm-continentaldivide.html

This downhill distance, however, wasn't going to be so easy. In the stretch
into Gallup, Google Maps showed no old Route 66; the route merged onto
I-40. In the past, we'd gone around such stretches on smaller roads, which
took us out of our way but which were far less dangerous. But this downhill stretch wasn't an option because any other roads would take us far into
the desert and create other risks, including our getting lost and having no
available power sources to recharge. We had no choice but to take the risk
and ride on the interstate.

Riding on the interstate in a city is risky, but riding in open country like
most of New Mexico, where traffic is doing 70–80 mph or more, is very

The Tour strategizes amid beautiful scenery and a formerly irrigated farm area at the Continental Divide.

dangerous. We had eighteen miles to cover before old Route 66 picked up again near Wingate. While I wanted to provide BenHop on the bike with some emergency-flasher cover, doing so posed the risk that drivers would clip the side of my car as they passed me while I was doing 20 mph on the berm. So I gave Ben a head start, and then I left the Continental Divide and passed him along the highway. This way, if he broke down or suddenly needed a battery, I would hopefully catch him as I drove by or only have a short distance to go back after I exited the interstate. This strategy worked pretty well, and we met up again near Wingate on Route 66, which went on into Gallup.

As we rode into Gallup, I got a call from "Jack" again. He was all set, having spent the morning hooking up the solar-power source he had to a connector that could charge my LEAF. He insisted that we meet him in town so he could use this opportunity for PR (public relations) to tout how solar power could be a reliable power source for an electric vehicle.

Interesting casino on the Navajo Indian reservation outside Gallup.
www.firerocknavajocasino.com

I understood his enthusiasm and the cause, so I agreed to meet him. In the interim, we came across the Fire Rock Navajo Casino on the way into town and took a quick look around. Then we went on into the city and met Jack near the Gallup Joint Utilities building in the center of town.

We met a small group of enthusiasts and exchanged pleasantries; they were very interested in our journey. Jack started checking everything out on the LEAF, and he was ready to charge. But for me, while I wanted to contribute, I wasn't comfortable with the idea of a homemade charge system, and my conversation with John Arnesen was running in the back of my mind. It just didn't seem smart to take the risk. While Jack would be performing a 30-amp 220v charge, the risk of a power surge remained. The risk put me in an awkward position, and finally I told Jack that I didn't want to take the risk of something bad happening. Jack didn't want to hear that and tried repeatedly to get me to change my mind. Thankfully, one of the other members of the Solar Power group stepped in and backed him down. I felt bad because I know he spent considerable time preparing for this opportunity, but at least some of the others understood my situation. Sorry, Jack!

After I made this decision, Benswing and I headed to the USA RV Park that we originally targeted. The managers there agreed we could charge up for a few bucks, and we were driven by one of the Solar Power folks back to Camille's Sidewalk Café, a restaurant near their office, for lunch. http://camillescafe.com/menu

There we had a great lunch with the Solar Power folks and the rest of

Recharging at the USA RV Park in Gallup, New Mexico.

the Tour who showed up. We gave an interview to the local PBS station in the back corner of the café, so at least the Solar Power folks could make the PR connection. To their credit, they were as committed to solar energy as some of the EV drivers we'd met.

After lunch, they took us back to the RV park, where we negotiated something like $20 for the power we used—a windfall profit for the park! With a 24-kw battery and even a high 14 cent/kw electric cost from the power company, the park's cost wasn't any more than $3.36—thus it made almost $17. Herein lies another of the challenges for electric vehicles; the average consumer has no idea that charging an electric car is very economical because electricity is fairly inexpensive.

We headed out of town on old Route 66 and met up with the rest of the Tour. But as we got just outside the city limits, Route 66 fed onto I-40, which we didn't want to travel. We followed Route 118 along the highway until it suddenly turned left under the freeway. We decided to follow it, but then it suddenly made a ninety-degree turn—right toward open desert. Then we had to backtrack to the Route 118 intersection with Route 66. So began an ongoing cat-and-mouse game of "road maze." Sometimes Route 66 was a side road, and we could use it. Sometimes it wasn't. When it wasn't, we tried to find parallel roads; sometimes they were well marked; sometimes they were not. Sometimes the road stayed on the same side of the interstate; sometimes it switched to the other side of the interstate and might be marked or might not be. Thus, we sometimes missed turns and had to double back. It was frustrating, to say the least.

We stuck together, however, and made steady progress to the state line.

Today was a pretty windy day (which we often experienced when winds blew from the west), so BenHop liked to have me take the lead and then follow me as I broke the wind for him. This sounds good, but it isn't as easy as it sounds. Ben liked to hug my rear bumper to get the best wind break, but, like most cars, my speed control won't work at 15–20 mph, so I had to try to maintain a steady speed by foot! Sometimes I found myself driving the car with a little too much power, so I'd pull away from Ben, and he'd get frustrated. At other times, I'd drive more slowly, and he'd want me to speed up. On more than one occasion, he ran into my bumper and put black rubber marks from his tire on my white bumper. Heaven forbid, but if I stopped too quickly, he would have been my rear trunk ornament!

Regardless of the wind, it was still a leisurely ride toward the state line, as we all held formation and took it easy. It was a good, warm day, with very little traffic: a perfect day to travel. And—a perfect day for an ambush!

As we approached the state line, I slowed the group down since I was the lead vehicle, and eventually stopped, seemingly in the middle of

BenHop rides my bumper for wind break as we push through easterly winds.

nowhere. Everyone was curious why we had stopped, and I stammered around with some small talk, while also giving Sean the signal to grab his weapon and giving Jonathan time to get his camera crew ready.

When they were ready, I told the group that the reason I had stopped was that I wanted to give them a proper "Welcome to the Wild, Wild West!" And with that, I grabbed my pump-action Super Soaker from the back of the LEAF and, along with Sean, began firing at will at all of the riders. We caught them by total surprise and no prepared defense.

Some just ducked. Others tried to grab their water bottles and fire back, but with minimal success. Others ran around the supply truck to get away. However, no one was spared: we even pumped some rounds into the documentary crew!

It wasn't pretty: Sean and I got everyone else pretty wet. Ultimately,

The calm before the shootout:
Rachel checks messages, and Dominique chats.

The NERF Scatter Blaster
pump-action Super Soaker
was the weapon of choice
for this battle!

most of them took off on their ride to get away from us. Victory was ours!

Eventually, the scooter riders collected themselves enough to try to reengage. As we continued down the road, I would have my window down, but after scooter riders would come up alongside me, I would wind my window up because they all tried to splatter me with their water bottles—but with little success. The wind didn't help their aim. However, when Susan came by, I noted she didn't have a water bottle, so I thought she wanted to talk. I kept the window down. That's when she spat a mouthful of water directly into my car and all over me. I must admit, she got me! But it was the only time. In fact, Sean and I were constantly looking over our shoulders for the rest of the trip, expecting payback for this one-sided shootout—but it never came.

Jonathan got his action and continued to film as we crossed the state line into Arizona. We triumphantly raised our fists in celebration as we each crossed the border. Seven states down, only two to go.

www.facebook.com/chiefyellowhorse

Shortly into Arizona we passed the Chief Yellowhorse Trading Post. We didn't get to stop, but the large cave with "Cliff Dwellings" certainly looked interesting.

The scenery was changing from reddish clay to more sandstone as we moved into Arizona. More mesas and flat desert land. It was beautiful in many ways, but still dry desert land with few natural resources to invite people to live in the area. It was one of those "nice place to visit, but I wouldn't want to live there" places.

Meanwhile, the frontage road was weaving first to the south side of the interstate, then to the north side, then to both sides—although the road on the south side suddenly ended a few miles outside of Sanders and the road on the north side suddenly deviated widely away from the interstate before it ended in Chambers, our destination. So knowing exactly which road to follow was like playing the lottery, because some roads had not yet been mapped by Google.

At the intersection where the frontage road switched from the south side of the interstate back to the north side, we were suddenly stopped by a surprising obstacle. It wasn't construction. It wasn't an accident. It was not a weather problem.

Instead, it was a small flock of sheep! And this small flock was tended by three dogs and a young boy on a bicycle. The boy and the dogs were herding the sheep from one section of the boy's parents' farm to another. To protect the boy's privacy, we didn't take his photo, because we didn't know if his parents would approve. This young man, perhaps ten years old, was doing a young man's job of moving the herd, and he instructed

Scenic landscape from eastern Arizona along old Route 66.

his dogs to help him do just that.

Surprisingly, our young sheep herder wasn't afraid of our large group, but we tried not to intimidate or interview him. We had a simple dialogue with him and then let him finish his task at hand.

On we moved as the sun started to set through more Indian reservation land.

We were getting close to our Chambers stop, but before we got there, we found yet another obstacle on our roadway. This time, it wasn't sheep. This time, it was a small river flowing freely over the road. It didn't look too deep, but currents can be deceiving, so the riders were unsure what to do. After most of us hesitated and just looked at one another, BenHop decided he'd go across the river and just pedaled right through, with no issue. So much for a sneaky and dangerous river.

Eventually, we all made it across the river without incident, just a little wet for the scooter and bicycle and motorcycle riders. On the other side, we posed for a classic silhouette photo. We had crossed another state line, another intersection, and now a creek. On we went.

A few miles down the road, we came to Chambers and ran into yet another obstacle! This time, it was road construction. A *lot* of it—and on

A small boy shepherding a flock of sheep captured
the Tour's attention along our route.

The documentary crew captures Susan
playing with the sheep, as Sean watches.

both sides of the interstate, as well as the intersection we needed to get to our motel on the other side of the freeway. It took some time to figure out how to get to it, and we ultimately backtracked to get there, but we finally got to our motel.

After we checked in and had dinner, there wasn't much to do. A few of us tried the pool for a bit, and then others tried to release their pent-up energy by climbing a decorative wall on the side of the motel. I'm not sure if the motel managers appreciated it or not, but there was a sense of excitement among the Tour members as Benswing successfully made it up—and over the wall. This is what happens when you've been on a long trip with few outlets for exercise or pressure release. Sometimes it's simple

There were plenty of roadside outlets for American Indian handmade items.

*BenHop pedals aggressively through the high water, and
then it's a question of who's next?*

*After a successful crossing, we posed for what turned out
to be an intriguing silhouette photo.*

things that lead to creativity and short-lived excitement.

It had been a long day, and I had to fully charge the LEAF at the motel. I did here at the Days Inn, as well, but it was close to midnight before I could unplug. At this point, it was "just another day," as they say.

CHAPTER 30

Day 27: Chambers to Winslow

"Highway Hopscotch." "Standin' on the Corner." "Ravin' for a Raven."

As we left the next morning, we knew our group was going to be split up a lot of the day. Our destination was Winslow, Arizona. Yes, the one from the Eagles song. Winslow was about ninety-five miles from Chambers and sat at a slightly lower elevation—4,800 feet. However, most of the Tour wanted to visit the Petrified Forest National Park, which lay nearby but required a significant detour from the Tour's path if we went through the whole park. So BenHop decided that a visit to the park would add too many miles to his day on the bike and that, therefore, he would stick to the more direct route to Winslow. I had been to the park before, so I agreed to support his battery needs while the others went to the forest.

Ben and I immediately had to deal with the interstate again, because the frontage road fed directly onto I-40 and there were no alternative routes. As before, BenHop and I played leapfrog, where I let Ben have a substantial time lead on me along the route, and then I would start driving, pass him, and then wait for him at the next major exit or off-ramp. Most of the time, this worked okay, but knowing where to exit to catch a frontage road that would stay close to the interstate left me totally reliant on the accuracy of Google Maps, and, quite frankly, that wasn't so good in 2013 for back roads. The roads sometimes seemed to have no rhyme or reason in how they started and ended or where they were even going.

But for the most part we did okay, swapping batteries out early so that Ben wouldn't run out of power without me around. Every once in a while, he'd be late, and I'd double back to find that he'd had a flat and was repairing the inner tube on the bike tire. The bigger concern was Ben getting thrown off the interstate. If the state highway patrol found him, it's hard to know what would happen. In fact, laws vary by state, but, generally, slow-moving vehicles, like bikes, are prohibited and considered a hazard to traffic on the interstate. He could get arrested. He might just get thrown off the interstate and have to find some other road to take, and who knows where that might go. And, to complicate matters, Ben's hadn't made any effort to replace his prepaid phone, so if I did lose him, well, I might be calling gas stations, hospitals, sheriffs, the highway patrol, and more, to figure out what happened to him. Not my idea of fun.

We had to ride the interstate for the first 30 miles, between both sides of the Petrified Forest National Park, and we had to ride until there was an exit with a frontage road. I passed Ben and decided to check out a roadside store called Stewart's Petrified Woodshop while I waited. Everybody in the area had petrified wood for sale, so I thought it might be interesting. Stewart's did have petrified wood, that's true, but what the store was exactly I'm still not sure.

I guess the dinosaur figurines surrounding the store were connected to the era of wood petrification, but I'm not sure how mannequins, old cars, a bus, and the other "stuff" were connected. The picnic table and the swing were a nice touch if you wanted stay a while, and at least the mannequins weren't inflatables. Still, it was a little weird. Stewart's just added a new line to help bolster sales. Wait for it…. How about ostriches and ostrich eggs? It all fits, right?

I guess this was the early cave man's version of bull riding.

*Be sure to check
out Stewart's Rock
Concert special:
www.petrifiedwood.com*

*The "interesting" collection
of bygone-era memorabilia
at Stewart's.*

After taking in everything Stewart's had to offer (and buying nothing), I headed back down the hill to the interstate to find BenHop. Once again, the side roads were confusing, and Ben ultimately decided to take his chances on the interstate since it was faster. So after another eighteen miles of leapfrog, we exited into the town of Holbrook, where we decided to stop for lunch. We found a small Mexican place on old Route 66 called El Rancho Restaurant and Motel. We ordered our food and then looked over a few interesting knickknacks the restaurant had while waiting for our food. The Mexican dishes were pretty good, and Ben and I relaxed, chatted with the owners, and talked about the day's sights.

Just up the street from the restaurant, we passed the Pow Wow Trading Post, another staple of old Route 66. Holbrook was an interesting town. It was obvious it was trying to uphold or invigorate the Route 66 mystique. There were numerous businesses, once thriving in the 1950s and 1960s, which were still alive today and touting tourists to "get their kicks on Route 66"! Yet Holbrook still had many more-modern buildings, as well. As we rode out of town along the old highway, we took note of a few of the historic landmarks still in place.

We turned in the center of town to return to the interstate and found

www.facebook.com/pages/Pow-Wow-Trading-Post

the Crossroads of Route 66, a small park and tribute to the history of the highway. It seemed everybody wanted to be connected to Route 66's nostalgia and potential tourist dollars. Some towns were clearly behind building a Route 66 tribute, and others were clearly not. Maybe that's typical of an aging population?

Ben was particularly disappointed because he had heard about this famous highway even in England, where he grew up. Yet major stretches of the highway were in disrepair, merged with the interstate, and, in some areas, were even nonexistent. Ultimately, it's hard to imagine sustained interest and business along old Route 66.

One more attention-getter as we approached the interstate was the WigWam Motel. Charles E. Lewis built this uniquely designed motel for tourists in 1950; his family still owns it today. Through the years, it has been modernized with new restrooms, air conditioners, and more, but the original fifteen individual sleeping units are still intact; there is also a museum, an office, and two small teepees, which used to be restrooms for the gas station in the front.

Giant mural of Holbrook history and sights at the Crossroads of Route 66.

Old West restaurants and cross-country travel hotels are typical themes for Route 66 businesses.

Classic cars from the 1950s and 1960s have been placed around the motel's "village" to add a nostalgic atmosphere; the Lewis family does encourage you to take a "step back in time." The Wigwam has had some notoriety over its years: most recently, it was the inspiration for the Cozy Cone Motel shown in the Pixar movie, *Cars*. We spent a half hour here looking around before returning to the road. Holbrook was an interesting change of pace, one we enjoyed.

Back to the grind, once again we had to play leapfrog on the interstate; there were no frontage roads. I let Ben get a lead and then passed him before exiting to check out the Geronimo Trading Post.

The Wigwam was once a motel highlight along Route 66 in Holbrook.

There are fifteen teepees for overnight stay at the WigWam.
www.sleepinawigwam.com/History/NewlyRenovated.html

Here I found more teepees, along with a lot of American Indian pottery and handmade crafts. There were a lot of these outlets near the reservations. I couldn't stay long, though, because, pretty soon, Ben came trucking down the interstate, passing me by. But I caught up to tell him to exit because there was now a frontage road we could take for a few miles.

On the interstate, off the interstate. On the interstate, off the interstate for the next twenty-five–thirty miles; it was like a game of highway hopscotch.

We finally drove into Winslow in late afternoon, and we quickly found ourselves at the corner. Not just any street corner, but THE street corner in Winslow, Arizona, made famous by the Eagles hit song "Take It Easy." As you might imagine, it's now commercialized. And if there is *anyone*

The Wigwam Motel offers drive-up entry to a personalized and cozy overnight stay.

who hasn't heard the song, you can check it out on YouTube: www.youtube.com/watch?v=LfeNhwnO8hw

I must admit, as a *huge* Eagles fan, I had a special feeling while standing on that silly street corner. I wanted to break out in song right then and there, but I decided to "Take It Easy," instead. And while there was a truck and a statue and a souvenir store, I saw no Girl in that red Flatbed Ford, and there wasn't any car slowing down for me (sigh). So Ben and I headed on to our evening event at La Posada Hotel across town.

I knew nothing about our event this evening, or about whatever La Posada was. Well, it turns out that the event wasn't much but La Posada was.

Once a spectacular Spanish-style hotel built along the Atchison, Topeka, and Santa Fe Railroad line by the Harvey Company, La Posada lives today as a fine hotel for railway or highway customers. It is a living and active museum with its collectibles, antiques, traditions, fine restaurant, and rooms suitable for celebrities.

Built in 1929 beside the ticket station for the railroad, the hotel,

More teepees and crafts at Geronimo Trading Post.
www.facebook.com/pages/Geronimos-Trading-Post

Commercialized and memorialized: STANDIN' ON THE CORNER: http://standinon-thecorner.com

A flatbed Ford and a bronze statue are fixtures at The Corner.

restaurant, and adjacent gardens were a convenient stop for railway riders. La Posada served thousands of customers for twenty years until the growth of autos and planes lead to a decline in rail traffic. It was converted to offices in 1957 and then closed until purchased and renovated in 1997 to save it.

Today, the hotel offers a number of appealing elements:

> ➤ Each rentable room is unique, including handmade pine beds, hand-woven Zapotec rugs, unique custom tile, six-foot cast iron tubs or whirlpool tubs, hand-painted murals, and more. The rooms are named for numerous celebrities, U.S. presidents, and more.

> ➤ An award-winning Turquoise Room restaurant.

> ➤ A Sunken Garden, Rose and Potager Garden, and Cottonwood Grove add color and living greenery to the desert environment.

> ➤ Historical artifact, art, and sculpture exhibit areas.

Ben gets a pic of the bronze statue "standin' on the corner."

My memorialized moment "STANDIN' ON THE CORNER"!

I met Daniel Lutzick, the general manager, who was intrigued by our Tour and joined us out in the circular drive entrance for our event. We spoke for some time about his efforts to revive the hotel and preserve its history. I think he's done an excellent job of it; it is a classy and classic facility to explore and stay in.

We had only one or two local electric-vehicle enthusiasts attend our event, but we were treated to having the Harvey Girls be there with us. This was a treat! The Harvey Girls were women who took waitressing jobs along the Atchison, Topeka, and Santa Fe Railroad line in its very early years, when there were no dining cars on trains. Over 100,000 young women from the Midwest took jobs along the rail line to send money back to their families through six-month contracts with The Harvey Company,

*BenHop and I arrive at the entrance to La Posada, which
also faces the Santa Fe Railway.*

which owned hotels and restaurants along the route.

The Harvey Girls took on wide-ranging tasks as part of their employment and were instrumental in the expansion of the West. They have been nationally and locally recognized for their contributions. True to their heritage, they were eager to check out our electric products of the future! These ladies were fun! They were also interesting as they shared their historical knowledge about the hotel.

After the event, we checked out the museums and gift shop. There were a lot of unique and interesting items from local American Indians and artists, as well as historical memorabilia. After we finished checking out the shop, it was time for dinner.

Our meal in the Turquoise Room was delicious! The meal was topped, though, by Ben Hopkins on this night. In the gift shop, Ben had spotted a puppet, yes, a puppet of a Black Raven. At least I think it was a Raven; it was a hairy, all-black bird, at least. Ben bought few souvenirs during our

www.laposada.org

Ben Hopkins entertains the Tour with his puppet-tales before we ride to the hotel.

trip, but, for whatever reason, he fell in love with this puppet. He bought it and proceeded to carry out humorous conversations with us through the Raven. Wacky, funny comments by the Raven.

Ben had us in stiches with his antics. (I think he was exposed to too much Benny Hill when he was growing up in Britain!) Even as we left and headed to the hotel, Ben was pedaling down the streets of Winslow, Arizona, with this Raven on his hand, talking to us. Ben was ravin' for a Raven. Kind of offbeat, like Edgar Allen Poe perhaps.

It had been a long day, again. Earlier, I had called the local Nissan dealer to confirm it had a Level 2 charger. It turns out it didn't even though Recargo's Internet site showed that the dealer did have a recharging location. Why didn't the dealer have one? You'll have to ask it. For me, it was another hotel charge night.

The Harvey Girls are ready to ride electric.

Join us for a memorable Tour photo.

Room # / Name		Amenities	Bed(s)	Rate
100	James Cagney		1 King	$129
101	Franklin D. Roosevelt	Patio Door	1 King	$129
102	Jimmy Stewart		2 Full	$129
104	John Huston		1 King	$129
106	Douglas Fairbanks		2 Full	$129
112	Jimmy Dolittle		1 King	$129
115	Grant Withers		1 King	$129
116	Roy Rogers	Full ADA Access	1 King	$129
200	Carole Lombard		1 King	$129
201	Amelia Earhart		2 Full	$129
202	John Wayne		2 Full	$129
203	Gene Autry		1 King	$119
204	Mary Pickford		1 King	$129
205	Shirley Temple		1 King	$129
206	Jane Russell		2 Full	$129
207	Will Rogers		1 King	$129
209	Victor Mature		1 King	$129
211	Jackie Gleason		1 King	$129
212	Dorothy Lamour		1 King	$129
214	Lionel Barrymore		2 Full	$129
215	Pola Negri	Walk-in Shower	1 King	$119
216	Isabella Greenway		1 King	$129
217	Frank Sinatra		1 King	$129
220	Omar Dooms		1 King	$129
223	Anne Morrow Lindbergh		2 Full	$139
227	Clark Gable	Standing Balcony	1 King	$129
229	Roddy McDowell		2 Full	$119
231	Gary Cooper		1 King	$119

Sample of the unique rooms at La Posada.

Sunset in Winslow en route to the hotel.

CHAPTER 31

Day 28: Winslow to Flagstaff

"See You Later, Crater."

"From Driving Courses to Beer For My Horses."

Today was going to be an easy day. Our distance to Flagstaff was less than seventy miles, although the Tour's drive would require a slow climb of about 2,000 feet. We could make it in roughly four hours if all went well. BenHop and I got off to an early start and followed old Route 66/Business 40 out of Winslow toward the interstate. We followed the road as it pulled up alongside the interstate and stayed on the south side of it. But, after about a half mile, the road suddenly turned due south toward the open desert and didn't merge with I-40 as we had expected. I've never encountered an interstate business route that didn't link with the interstate, but this one didn't. We retreated to the last major crossroad, crossed to the other side of the highway, and followed the frontage road.

The frontage road followed alongside the interstate, and we were moving briskly along at 20 mph. That is, until the road suddenly just-- ended. No warning. No signs. It just stopped about two and a-half miles from where we had gotten on it. It turned out that the frontage road picked up again about eleven miles further up the interstate, but that did us no good here.

BenHop and I discussed the problem and decided it made no sense

We hoisted BenHop's bike over this fence to avoid backtracking.

for him to pedal all the way back to the frontage road and then onto the freeway, adding five miles to his day, so we walked his bike through the desert terrain to the four and a-half foot high fence separating us from the highway, and he and I hoisted the bike over the fence. Ben found a way to get over it, as well (though I suspect there are still a couple of sagging links in that fence!), and he continued traveling on the side of the interstate. Meanwhile, I drove back to the entrance of the frontage road and got onto the highway to catch up. Maybe today wasn't going to be such an easy day, after all.

So, once again, Ben and I played leapfrog on the interstate. I would hit the exit or rest area ramps and guide him on whether to keep going or to exit for a side road. Our game of leapfrog also meant I had extra time to kill along the way: to plan, to sightsee, to explore, to nap, to rock out, and more.

Ben and I were intrigued by the next major exit, which was for Meteor Crater, and we wanted to detour to it. Our detour would be difficult, however, because we would have to drive over six miles to the crater, which was south of the interstate; this meant twelve miles of additional pedaling for Ben on that day. But he decided a visit was worth the extra effort, and off we went over Meteor Crater Road, a slight upgrade, though straight, until we neared the crater.

In about twenty-five minutes, Ben and I made it to the Visitor Center and soon climbed to the rim of the crater and looked down inside it. What we saw was amazing! Meteor Crater is nearly one mile across, nearly two and-half miles in circumference, and more than 550 feet deep.

It is an awesome sight; one which makes you think about the enormous

Exploring a rest area rock garden, I asked
someone to document the moment.

impact this meteor must have had (estimated as about 20 million tons of TNT!) to have pushed so much earth upward to form the crater. In relative terms, it really wasn't that long ago when the meteor fell—i.e., 50,000 years! Dang, I just missed it.

The experience reminded me of my time at the Grand Canyon. The Grand Canyon always makes me think of just how small my time on Earth is relative to what it took to make the Grand Canyon. The Meteor Crater made me think about just how small I am in relation to the size of the universe. Pretty small. Almost like saying that my lifetime wouldn't measure up to be a pimple on the Earth's butt, I digress My sight of the 550 feet of the Meteor Crater was a true moment-in-time retrospection.

Ben and I elected not to walk to the crater floor, figuring we might be just a tad winded by the time we got back to the top. But we did give a mental salute to those that had done so: it was a long way down there and back!

I would like to return someday to do the rim tour and to go down to the three observation platforms in the crater. There are also outdoor observation trails, a wide screen movie theater, an Interactive Discovery Center, a unique gift and rock shop, and an Astronaut Memorial Park at the Visitor Center that would be worth more time as well. "Next time."

Having gotten the "Hole" experience, we knew it was time to move on.

We returned to the parking lot just as it appeared some rain clouds were forming and to find that Ben had a flat tire. So right there in the parking lot, we broke out the inner tube patches, the glue, and the electric

*From top, you can see the enormous impact of
a meteor traveling 26,000 mph.
http://meteorcrater.com*

*Observation platforms allow you to view the crater
floor and the "secret" display in the middle.*

pump (powered by my cigarette lighter outlet), and Ben got to work.

Having cycled across multiple continents and countries, Ben had inner tube repair down to a science. And while I provided support, I thought it better to try to stay out of his way so he could repair *his* bike the way *he* wanted.

When he was focused, Ben could be very efficient in tearing down the

Ben Hopkins and I take in the awe of Meteor Crater; that's one big hole! A Visitor Center wall provides a photo frame of the flat surrounding terrain near the crater.

bike, repairing the tire, and reassembling the bike. He wasn't just happy with a repair, however; he usually wanted to know what caused the flat so it wouldn't reoccur with the same foreign material in the tire. But he was very conscious about how the bike itself performed with the sustained grueling grind of multihour usage, the weight it was carrying, the heat, and even the number of bumps and turns it had to endure for a ride of this magnitude.

For this teardown, Ben had some concerns about the shifting mechanism and the spokes, but he ultimately just repaired the tube so we could be on our way. The storm clouds were gathering.

With storms that blew up to the south of us and just to the east of us,

Rain looks likely as we fix Ben's flat tire at Meteor Crater.

I don't know how, but we managed to avoid getting drenched—well, rather, Ben didn't. We did get some light rain, but not much, and we soon had returned to the highway. Once again, however, we were forced to ride on the side of the interstate for another thirty miles until we reached Flagstaff.

There were roads running alongside the interstate, but

1) they often took sudden perpendicular turns to places unknown, and

2) *none* of them connected one exit ramp to the next!

In all my lifetime of driving, I don't think I've ever encountered that before, and hope I don't again—crazy!

We began our leapfrog habit again, and lo and behold, as I passed Ben and drove to the next exit, whom should I find? The scooteristas! Susan and Dominique hadn't stopped at the Crater as we had and were just catching up to us since they had started later in the morning. They would have beaten us to the event in Flagstaff, but we were all pretty close together this time. I stopped on the exit ramp to wait for Ben (to pick up Old Route 66 again) and took a couple of snapshots of Dominique at the same time. The photos show how close the riders were to the big trucks doing 40–50 mph faster and which could literally blow the scooteristas and the bicyclists off the road!

Ben soon came along, and we headed to an event near downtown Flagstaff at Lumberyard Brewing Company. It would be a consumer event where we hoped to encourage patrons to ride our electric vehicles and to talk electric.

Located across from the old steam train station and part of the Historic Railroad District of Flagstaff, this former lumberyard dated back to the 1890s! Because the nearby foresting ended in the 1990s, the old lumberyard was bought and then converted to a brewery and restaurant—and was now family owned.

*BenHop pedals hard to outrace the rain, and we head
to the interstate: See you later Crater!*

When we first got to Lumberyard Brewery, it was a little awkward. The owner, Evan Hanseth, wasn't aware we were coming. His daughter Julie sets up all their events and didn't tell him. Oops! But after a little explanation on what we were doing, Evan bought in, and he not only set us up for fun but was the first one to ride one of the electric bikes. His ride was a little shaky at first, much to Julie's delight, but he got the hang of it and found it to be a pleasant ride. After that, we pretty much could do whatever we wanted at Lumberyard Brewing!

Soon it was announced to the bar and restaurant patrons over the loudspeaker that the Ride the Future Tour group was here and that anyone

*If you want a thrill sometime, try riding a bike or a scooter alongside
an interstate with eighteen-wheelers passing by.
And if you're a Susan daredevil, put a purse between your legs and
make cell phone calls!*

*Dominique makes it to the Flagstaff exit ramp safely,
after avoiding road trash.*

who wanted to ride an electric bike or scooter should head outside to the parking lot. And we did get curious folks who wanted to talk turkey about electric and ride the bikes and scooters.

We even had a competitive electric-bike owner show up. He rode a Stromer from Switzerland, and I decided to try it out. I was very impressed with its power in comparison with that of Ben's A2B bike, although I knew Ben's was a more midrange bike and the Stromer was high end. I could see myself owning one of these babies! The Stromer was very smooth and a lot of fun.

For the next hour, or longer, we were busy with consumers test riding and talking electric. This was a crowd that was proenvironment, so it wasn't a sell job as much as it was a "give me the details" set of conversations.

Meanwhile, *Green Lightning* had been given a primo seat for display amid the brewery vats of beer. I'd noticed that she'd been a little sluggish today, so I decided to recharge with a new fuel this time, from the vats! Just like the Toby Keith and Willie Nelson song, I decided the best way to get *Green Lightning* ready for the climb to the Grand Canyon tomorrow was to give her a boost of "Beer For My Horses!" Everyone thought this was a great idea!

This proved to be a good conversation starter with the consumers walking by. It was a good laugh, and I appreciated Evan's flexibility of allowing us to basically become part of their facility. The Hanseths were very nice people, and the evening was a lot of fun. We later all sat down to dinner and had some great American food and drinks. It was a fun estab-lishment that was actively part of the community, with regular nights for

*Sean instructs Evan to use the throttle lightly and pedal, as
daughter Julie takes photos.*
www.facebook.com/lumberyardbrewingcompany

*Attendees try
riding a scooter
and bikes, and
are impressed.*

Bingo, Trivia, Boys Night Out, Ladies Beer Club, Country Dancing, and Mimosa/Bloody Mary drinks. Other special events were also added at times—like us!

When everyone was finally worn out, we headed to the motel. It was time to recharge the LEAF again, and I wanted to use the local Nissan dealer for convenience, but Planet Nissan was not a participating LEAF dealer! This was odd to me, because this is an area concerned about the effects of global warming and the environmental impact on the Grand Canyon just to the north! And there's no LEAF market?

Anyway, once again, I'd have to charge at the motel. This evening would be cool, though, and I could open the window. So at least I wouldn't have to sweat all night while the LEAF charged. It turned out that this was a good day, after all.

Recharging Green Lightning *at Lumberyard
Brewing: "Beer For My Horses!"*
www.lumberyardbrewingcompany.com

Listen To Willy Nelson and Toby Keith sing: "Beer For My Horses"
www.youtube.com/watch?v=o1JOFhfoAD4

*Note: Scenes from the video "Beer for My Horses" were shot at the Mine Shaft Tavern in
Madrid, NM!*

CHAPTER 32

Day 29: Flagstaff to Grand Canyon Village

"Road Rage."

"Thou Shalt Not Steal From Little Ladies of Age."

It was bound to happen sooner or later. And I guess that after yesterday's issues, I should have expected it. We had gathered early this morning to prepare for our group ride to the Grand Canyon. As I had done every day since the start of the trip, I had made copies of the route map for the day, which included the written directions from Google Maps. That task wasn't always easy to

1. confirm the route details for the next day and make any needed changes,

2. get the exact size map *and* the written directions together, and

3. load both onto a flash drive, before

4. going to the motel office and downloading it all for print—which in itself was sometimes a challenge.

It was a pain-in-the-ass task which I did every night even when I was tired after a long day. But I did it so everyone had the info before they headed out each morning. It was especially important since the group members were often split up and had to find their way without a map at times.

For most of the trip, this process had worked pretty well. But in New Mexico and Arizona, Google's mapping wasn't always clear or complete; thus, sometimes, we got confused. And inaccurate road signage could further complicate our confusion.

The issue came up for discussion this morning, as I handed out the maps. Evidently, the riders had not gotten my message yesterday after BenHop and I got confused. Whether they read my message or not, I don't know. But the riders had also gotten confused and lost and had to retrace their paths a couple of times.

I explained that I always tried to take the best route and stay off of the interstate. Furthermore, I explained that BenHop and I had encountered a road that the map had showed going through but in reality had just ended. Dominique was becoming frustrated and stated that what we were doing wasn't working and that we needed to take other routes. But while I tried again to explain that the Google Maps weren't always clear and that I thought these were the best roads, my explanation seemed to fall on deaf ears.

Well, right or wrong, quite frankly Dominique's criticism pissed me off. Had it come from Benswing, who regularly used the maps to navigate, or even Susan who helped me identify our route initially, I would have accepted the criticism better, but I knew Dominique never had involved herself in any of the route planning. I doubt she'd ever looked at Google Maps other than the copy I gave her each morning. She didn't even seem to appreciate the work I was doing to make it easier for each member of our Tour to know where to go each day. For her to gripe about the route selection simply rubbed me the wrong way. I knew it was wrong, but, from that day on, I never gave anyone directions other than BenHop, because I knew he didn't have access to a laptop. I figured the rest of our group could decide what routes they wanted to take on their own. I somewhat regret that decision, but, ironically, no one ever asked me to keep handing out maps either.

If there was going to be an issue about which routes to take, it certainly wasn't going to happen in the next couple of days, because our route to the Grand Canyon was a deviation away from I-40 and there was only one road north to the Grand Canyon. We'd use that same road when we returned tomorrow; i.e., Arizona Route 180, which merged with Arizona Route 64, which took us to the canyon. And the road was a flat eighty

Dominique and BenHop check their vehicles; "only one personal carry-on allowed, Dominique!"

A local electric-vehicle enthusiast joined us for this leg of the journey on an A2B bike.

miles all the way to the rim.

As we prepared to embark for the day, it was hard not to notice the significant change in altitude, temperature, and terrain. The area was full of large pine and deciduous trees. It was the nicest weather we'd had all trip—the air refreshing and pleasant. It certainly beat the heat and rain we had dealt with so far, and it was a beautiful day to travel to one of our most magnificent national parks!

Route 180 out of Flagstaff was a divided highway when we started out of town. It was wide and flat, and we could ride with ease. I provided a little cover from the rear for the riders by using my emergency flashers. Our initial stop would be where Route 180 and Route 64 met, which lay about sixty miles from Flagstaff. Great weather and flat road meant we made good time getting there, stopping only long enough to swap out the bike batteries along the way.

The group is "ready to rumble" to Grand Canyon Village.

*The tall pine trees lining our route were a stark
contrast to the earlier Arizona desert.*

Open plains gave us sweeping views during our ride.

The road narrowed as we ventured further out of the city. The trees were plentiful, and the scenery got more picturesque. We then broke out into some open-plain areas, which were awesome in a different kind of way.

We arrived at the junction of Arizona Route 64 at around 12:30 p.m. and stopped at the Valle Travel Stop for lunch and the restrooms. This wasn't an in-house stop; in fact, we all just bought some snacks or sandwiches and drinks and sat at tables outside the store to chat. It so happened that there were a few custom cars positioned around the parking lot, so we mingled around for a look. The cars added a little color and pizzazz to the place, and we talked to a couple of owners for a few minutes. I don't think electric vehicles were on their radar, though.

Soon we were on our way again for the remaining twenty-nine miles to the village. Things were pretty calm until we got about halfway there or more, and suddenly the traffic stopped ahead of us. We soon saw the reason: a large elk had crossed the road and was eating on the roadside. We took a couple of pictures and moved on.

As we got closer to the Grand Canyon National Park entrance, I got a call from Tiffany, the gal that Jonathan had recruited to help Susan with our press and events. Now that we were getting closer to accomplishment, it was even more important, and Susan simply didn't have the time

*BenHop and our rider guest pause for a few
minutes after swapping out batteries.*

A scenic, leisurely ride along Arizona Route 180.

The classic cars were a supplement to our lunch stop.

to arrange for events, to gather local support, and to make sure various media outlets were present to cover our Tour. Tiffany was our contact, and she tried exceptionally hard to stay ahead of us and gather local support; she was good at what she did, and we needed her because our Tour had no real resources or local credibility with the press.

Tiffany was returning my call because I was trying to find out where I could get 220v power at the campground Susan had booked for us to

Custom rods at Valle Travel Stop.
https://grandcanyontravelstop.com

Elk are not native to northern Arizona but do live among the Ponderosa Pines.

stay that night. What I learned was that no reservation had been made at a motel or a campground. Strike 1. Tiffany had now made a reservation at the only available sites in a primitive campground—which meant there was *no* electric power anywhere in the campground. Strike 2. Now, I'm all for a Kumbaya gathering for bonding. It's healthy occasionally for morale and such. But to book a campground site for an *all electric vehicle tour* in a location with *no electrical outlets* seems a bit extreme. I mean, what were we doing?

So when I asked where I could recharge the LEAF overnight, Tiffany told me I'd have to find some place in the area to recharge. Strike 3. At this point, I came unglued. In certain "unfriendly terms," I told Tiffany that wasn't acceptable and that she needed to find me a place with 220v power. I knew there were no public chargers in the area, and I wasn't sure that other campgrounds in the area had RV outlets with 220v power.

To Tiffany's credit, she handled my "unfriendly terms" well—though she teasingly reminds me of that moment every time we talk today, noting that she cried afterward. (For that I have apologized!) But I was at real risk of not being able to recharge the LEAF overnight or having only a partial charge because I could use only a 110v outlet. So Tiffany went searching while we continued into the Grand Canyon National Park and approached the South Rim, where Jonathan wanted to film the group.

About this time, Tiffany called back. She had gotten resourceful and called the Grand Canyon National Park Service to explain the situation. Someone there transferred her call to the Sustainability Group, who was intrigued with our cause and did two things:

1) they had her tell me to go to the campground where the summer concession workers lived in trailers and tents and to recharge at an open campsite there, but I had to be out by dark;

2) they asked a Sustainability Group leader, David Perkins, to meet us at the Rim to talk about the Canyon and hear our story.

Kudos to Tiffany! This was *fabulous* news! I now had a power source to recharge the LEAF. So I drove over to the campground and met Mr. Perkins, who told me to pick any open campsite, which I did, plugged in, and we headed back to the South Rim to get in on the documentary filming. Meanwhile, about this time, some dangerous thunderstorms had formed. And at the park's elevation, lightning was close and the thunder

Green Lightning takes a charge from a campsite in the Grand Canyon National Park.

Enjoying a rainbow over the Grand Canyon.

extremely loud. Storms were around us, as well as out in the canyon itself, so the filming took place and the pictures were taken in between storms. At one point, I remember being stuck under a trolley-stop cover as rain poured. Nevertheless, since the rain was sporadic, we were able to get views of the Grand Canyon and all of its splendor.

With the rain, it did get noticeably cooler at the South Rim. But for many of the Tour group, this marked their first time at the Grand Canyon, and they were in awe of it. You can't help being in awe; the views are simply breathtaking.

One of the concerns Mr. Perkins noted that the National Park Service had was the impact that air pollution from LA and the West Coast was having on the coloring of the Grand Canyon rocks; air pollution could permanently alter the walls. Some of the park service's efforts are focused on trying to minimize the impact of pollution by reducing the sheer volume of vehicles that get within the national park grounds; ecofriendly trains, buses, and other modes of public transportation are included in its plans.

The Grand Canyon Sustainability Group has many practices in place, including an aggressive recycling program in both public and residential areas within the park; a water reclamation facility that converts wastewater into water safe to use for irrigation and toilets; a composting program for mule waste from NPS and concessionaire operations; and a shuttle bus fleet that runs on clean-burning compressed natural gas. In addition, all recently constructed buildings in the park meet LEED (Leadership in Energy and Environmental Design) standards, and all future construction will be designed to achieve the highest LEED ratings. The Sustainability Group even rewards and acknowledges those concessionaires with the most creative or effective effort for sustainability each year! Now that's a full-team approach to sustainability!

In 2010, the Grand Canyon also became a Climate Friendly Park (CFP), meaning it has made a commitment to become an environmental leader by lowering its greenhouse gas emissions and teaching others how to limit their own impacts on the environment. So our little electric parade was right up its alley. We were visiting the park, enjoying our time, and without carbon emissions that would negatively impact the park's surroundings!

The rain finally stopped, and we wanted to get to our campsite for dinner, so the riders and documentary crew headed in that direction, while Sean, Benswing, and I headed back to the concessionaire campground to pick up the LEAF, which would be fully charged by now.

We were chatting as we drove up the gravel alley to the campsite, but just as we got close, I noticed someone scurrying away from the LEAF and back over to the campsite next to it. I thought this was odd, and I asked the guys to wait until I made sure everything was okay. I was hoping someone was just curious about our travels, but one can't be too sure.

I jumped out of the supply truck and went over to the charge indicator to see if the LEAF was fully charged. It wasn't. What? Why not? I checked the charger; it was dark. I checked the outlet; the charger had been *unplugged*! "What the hell?" I thought. I plugged it back in, and it worked!

Obviously someone, possibly the person who scurried away when we drove up, had unplugged the charger. Problem! It meant I had to wait another two or three hours to get the LEAF fully charged! But who would

Dave Perkins from the National Park Service's
Sustainability Group talks to us about the park's programs.

A Ride the Future Tour group photo on a rainy day at the Grand Canyon.

*Jonathan talks
to Susan about a
movie shot.*

do this? And why?

I soon had my answer. A rather tall and husky fellow from a trailer across the alley came sauntering over and proceeded to ask us who we were, what we were doing, who gave us permission to use that campsite, and more. I really didn't care for his third-degree questioning or his approach, but we answered his questions. About this time, our little scurrying friend came out from the tent next door; *she* was a lady probably in her seventies or eighties, a little hunched over, and matching the vagrant image that her campsite reflected. I don't recall her name, but let's call her "Rudey."

When we further explained to our inquisitor that someone had unplugged our connection and that our car was not recharged, "Rudey" informed us that she was the one that had unplugged the charger. And

Dangerous storm clouds were very close to our observation area.
www.nps.gov/grca

when I asked why, she stated that she had come out to see what we had done after we left and saw the shiny round disk on the electric meter spinning out of control, so she unplugged the charger to make the disk stop spinning. "What? Really?" Rudey went on to say that she didn't want to pay for all that electricity being used. "What?"

So, for about the next fifteen or twenty minutes, Benswing and I spent time trying to figure this out and educating Rudey and her neighbor about the costs of electric-vehicle charging. We told them we had permission from the National Park Service to use any campsite, but according to the two of them, Rudey had the rights to our campsite. She had recently switched to the one next to it. "Why?" I don't know, and why she thought she still had a cost tie to "my" campsite, I still don't understand. But she firmly believed that I was stealing electricity from her. We finally explained that even if that *were* true, the cost would have been less than $5!

It was about this time that I heard a vehicle drive up. Yep, seems our friendly neighbor had called the park rangers before he came over to talk

*The campsite next to the LEAF had scattered belongings,
an ATV, and a "friendly" neighbor.*

to us. So *now* we had to explain all this again. But before we did, I got the third degree from the rangers, including a request for my driver's license, my car registration—which, of course, was Nissan's and not mine—the Borrowed Vehicle Agreement, and other documentation. They ran the license plate to see if it was stolen. I mean, these dudes took this very seriously, and, at one point, I thought I might even get arrested!

Thankfully, Benswing was there and helped explain the kilowatt usage of electric cars and low cost implications. I even offered to pay Rudey $10 to make her whole. But, no, she wouldn't take it. "A-r-r-r-g-g-h!" We explained that we had no idea the campsite wasn't available because it appeared to be unoccupied. Furthermore, the Sustainability Group had told us to pick any open site.

Finally, after listening to our story, and finding out that my car wasn't stolen and that I had no "wants" or "warrants" for my arrest and that thus likely I wasn't a wanted fugitive, the rangers decided to let me go. They made sure Rudey was comfortable with our explanation, and I was advised to finish my charge and be on my way and not stay overnight. "Happy To!" With that, they drove off.

Rudey's friendly male neighbor then went home—and probably immediately applied to be head of the Concessionaire Campground Neighborhood Watch Committee, turned out a Buford Pusser "big stick" on his lathe, and proudly patrolled the area the rest of the summer, looking for other energy and noise abusers. If he didn't, I'm sure he popped a beer, put on a robe, and turned on his TV to watch *The Jerry Springer Show.*

At this point, I was really beat. Benswing decided to go over to the campground, but I had the luxury of standing and sitting around for the next two–three hours while the LEAF charged up. By now, Rudey was feeling guilty, and she wanted to chitchat. I don't remember much of the conversation other than

A) she wouldn't take the $10 which I repeatedly offered, and

B) she was the proud aunt of Olympic swimmer Greg Louganis.

I do wonder if Greg visited her campsite often? Nevertheless, Rudey and I parted as friends, and I didn't get arrested.

Greg, if you're reading this, let the record state that I never tried to steal electricity from your aunt!

It was around 9:30 p.m. when I finally finished charging and met the group at the campground. It was dark by now, but Sean was the chef du jour for our camping night, and he was cooking ribs; they were delicious! We had another Kumbaya moment at the no-luxuries campground, and we did have a good time swapping stories. I had one to tell that night, and for years to come.

Ironically, after all this, Susan had found two motel rooms to charge the bike and the scooter batteries, and she offered to let me sleep there instead of the tents that night. I accepted.

Grand Canyon beauty.

Ben Hopkins: This guy made a lot of noise and was looking for you. He's a relative of your "raven" puppet, and he's pissed!

CHAPTER 33

Day 30: Grand Canyon Village to Seligman

"First Aid Fix."

"Getting Our Kicks on Route 66."

On an adventure such as a cross-country trip with people you don't know, you learn a lot—not only about the attributes of each person but also about what you would do the next time differently—assuming you want a next time! That was true on this trip, too. And the first thing I would change would be the schedule so that there was some time to rest along the way. You know, like time to smell the roses, so to speak—or, like our morning today, to be able to smell the fresh air and the mule dung from the riders headed down the Bright Angel Trail in the Grand Canyon.

Our schedule was tough—almost grueling because there were no breaks along the way. Every day, we got up, prepared, rode all day long, attended an event, had dinner, went to bed, and did it all over again the next day. And even though we stopped at some great places like the Grand Canyon and wanted to explore, or perhaps wanted to talk to more consumers about alternative fuels and ideas, our schedule didn't allow it. Nor did it allow us the chance to just catch our breath for a day, and, toward the end of the trip, I believe our hectic schedule affected us all.

While it would have been preferable to enjoy the next beautiful day seeing more of the Grand Canyon, we packed up instead and headed back down Arizona Route 64 toward Interstate 40 again. And, today, Rachel decided to ride with BenHop on a bike and we had a guest rider/enthusiast, as well. Let's call him "Mike." "Mike" was a good friend of Dominique and Susan, and he had shown up the night before at our Kumbaya campfire for eats and drinks. A nice guy, Mike was an athlete in very good shape and had a fun personality, which fit in easily.

Our ride would take us due south from the Grand Canyon to Williams, Arizona, where we would pick up I-40 again, have lunch, and then head west to Seligman. The total distance was around 102 miles, which makes for a pretty full day. The road to Williams was good with pretty countryside and not a lot of traffic, although it was a two-lane highway. The frontage road along I-40-well, who knew actually what that would be like, given our recent experience?

We headed down Route 64, and I followed the bicyclists for support. As was the norm, the scooters and the supply truck would be behind us and Benswing on the motorcycle somewhere in between.

After about one and a-half hours, the three cyclists kept me busy providing batteries each time they ran out of power. With another beautiful day, we made good time. Williams, Arizona, was our stop for lunch, and it was about sixty miles from the Grand Canyon. After I had replaced the second battery for each rider, I decided to go ahead to scope out a place to eat. I was also hoping to get a peek at the Grand Canyon Railway steam engine, which originates from Williams and makes treks to the South Rim of the canyon.

I had only gotten about five miles ahead of them when I got a surprise call from a number I didn't recognize. It turned out to be Mike. He told me that Rachel had just had an accident, and he asked me to come back to them with the first aid kit. Apparently she was okay, but had taken a pretty

Ben, Mike, and Rachel take a break to swap out batteries.

hard fall. I immediately pulled a U-turn in the middle of the highway and raced back to where they were. About this time, Benswing had come along and was also there to help.

I left the car part way on the road and part way on the shoulder, with the emergency flashers on to provide some cover for us from traffic. Rachel was up and walking around and appeared to be okay, but a bit shaken. She had gone down hard, because she had been just a bit too close to the berm, which had some loose gravel on it. The berm wasn't very wide, and just past the berm was a steep downhill slope of gravel leading to an open field. Rachel had been riding on the berm, and, for whatever reason, she got on the right edge of it when it suddenly just gave way on her, spilling her onto the gravel as the bike slid away from her.

When I first saw Rachel, she had numerous bloody scrapes on both her hands and knees. Fortunately, she had not hit her head or broken any bones or suffered a more serious injury. She was in a little pain, but she still kept her sense of humor about the accident. I broke out the first aid kit, and BenHop and Benswing started reading through some written instructions to figure out what would be the best first aid items to use. They found dressing and some salve we could use, but we didn't have anything to clean the wounds. And there was no place anywhere close to us to get any alcohol or soap. We didn't even have any water, or did we?

I remembered I had a fully loaded Super Soaker ready and waiting for emergency use in the LEAF. It would work! I grabbed it, and as we poured the Super Soaker water onto her hands to clean the wounds, we also

BenHop and Benswing look for the right first aid in the kit, while Mike documents the event.

BenHop and Benswing work to give first aid to Rachel after a hard fall.

Cleaning Rachel's injuries before we wrap her up with gauze.

picked gravel out of them. When we got them pretty clean, Mike jumped in, applied some salve, and used the gauze to wrap her hands.

Through a group effort, we got Rachel fixed up. We all felt bad about her injuries, particularly Ben Hopkins, who somehow thought her accident was partially his fault. But it wasn't. It was a simple accident that could have happened to anyone. Because this was only our second incident of any kind some 2,000 miles into our trip, we had to be thankful. I always dreaded the thought of hearing we had a rider down somewhere.

Hearing Rachel had an accident had sent chills down my spine. It's the last thing any of us wanted to hear.

Rachel was a real trooper. She never complained. Sean had gotten the word and rushed over with the supply truck to where we were. We loaded Rachel's bike into the supply truck, and Rachel, Benswing, and I drove on and into Williams to find a medical facility. We found one, and while Benswing waited with her and she did paperwork and waited to be seen by a doctor, I drove over to a nearby RV park, where I could recharge the LEAF. It was a pretty basic campground, but the power was free, so I plugged in and headed back to find Rachel and Benswing.

"This won't hurt. We promise."

In-city campground in Williams, Arizona and
my portable charger hookup for "Green Lightning".

The RV park was several blocks from the medical clinic, and I had to hoof it. No big deal; I didn't mind the walk. What I did mind, however, was the cracked glass face on my cell phone after I accidentally dropped it on the city sidewalk during my little hike. Thereafter, I had to look at a splintered picture of everything I did on the cell phone. That sucked. Every day.

When I got to the clinic, Rachel had just been seen by a doctor—the delay partly due to her Australian insurance card not working too well in Williams, Arizona. She had bandages on her hands and knees but seemed to be in good spirits. She was good to go after being checked out, and we all headed over to a nearby drinking hole, Pancho's Mexican Restaurant, and met the rest of the Tour for lunch.

Lunch was an opportunity for Rachel to explain what had gone wrong during the ride and to show her bandaged hands. She never complained, and we were all able to laugh about the whole incident.

Later, a few of us took a walk around town, checking out some of the businesses. Williams is another of the Route 66-promoting towns, so there were numerous themed shops and restaurants.

I then headed back to the LEAF, but not before stopping at the Williams train station. Here I found the Grand Canyon Railway Hotel and the historic Grand Canyon Railway steam engine used for pulling tourists from Williams to the South Rim. It is in great shape and a reminder of typical transportation used by millions of Americans during the growth period of Route 66. The steam engine is only used on weekends (the bulk of the work now done by diesel engines) but has been converted to run on

*Pancho's was our lunch venue, where Rachel showed
everyone her injuries and told her story.*

recycled French fry oil! The railway itself keeps over *50,000* cars from traveling into the Grand Canyon National Park each year!

I walked back to the campground and unplugged the LEAF. Soon we were ready to resume our travels, which required us to travel another twenty-nine miles on Interstate 40. The traffic wasn't heavy, and we made good time across the desert.

As we got about twenty miles from Seligman, we were finally able to exit onto a resumed portion of old Route 66, which leads into the city itself. At the city limits, we were greeted by a sign telling us that Seligman was the birthplace of Route 66; I learned this after I discovered that Seligman had convinced the state of Arizona to make Route 66 a historic highway, which, I also discovered, led to other states taking similar actions.

As we rode into the center of town, it became obvious that Seligman was a Route 66 promoter. Here we found business after business with a 1950s theme—although a few current movies, like *Cars*, with throwback

Downtown Williams offers lots of Route 66-themed restaurants.

*The Grand Canyon Railway steam locomotive runs only on weekends
these days to the canyon's South Rim.*

themes were woven in! These themes were happy themes, images of fun
and humor.

I took a walk around just to check out how the town has tried to preserve its Route 66 image of the 1950s. Here was a place where Route 66
signage was *everywhere*! Old cars, trucks, signs, toys, pictures, tools, appliances—you name it. This town relies on the image of Route 66. I suppose
that in a small town in the middle of the Arizona desert, there are not
many alternative business opportunities, so trying to get passersby to stop
and spend a few dollars to relive the 1950s is as good as any.

After checking out the locale, I headed back to the Romney Motel, our
stop for the night, to charge the LEAF. The hookup in the room was going
to be difficult, so the manager connected me to an outside 220v outlet
near the office. I didn't really like the setup because the outlet looked like
something that might blow a fuse as soon as I turned on the charger or

that would lose power in the middle of the night, but I didn't have much choice. On the other hand, the makeshift setup matched the condition of the motel, so, in an ironic sort of way, I guess the setup was appropriate.

We were greeted by this Route 66 birthplace sign and a 1950s drive-in burger joint in Seligman.

Gas stations kept their retro 1950s look on Route 66 in Seligman.

This general store featured everything from Mater (in the movie Cars) to merchandise and humorous signage. www.seligmansundries.com

Distance signs and Tow Mater greet you at Seligman Sundries.

225

Seligman features numerous Route 66 businesses of the 1950s.

The Romney Motel: once an overnight staple on Route 66, but definitely not Mitt Romney's.

The mannequins in this aging Buick at a car wash are a nice touch!

CHAPTER 34

Day 31: Seligman to Kingman

"Someone has a Giganticus Headicus."

We were now one month into our adventure, with only two weeks remaining. Daily routines were pretty much just that—routine. We all knew what we had to do each day, and we did it, individually and as a group. Most of the time, we held group dinners in the evening, but, sometimes, you just had to get away.

Today was another routine desert day. Our distance was just under ninety miles for the day, and, even better, we would be descending almost 2,000 feet—from the 5,200 feet above sea level in Seligman down to 3,300 feet. Downhill was easy for the bikers, and easy for me since the LEAF uses less power. Perhaps more important, Route 66 took its own path out of Seligman, diverging from Interstate 40 and making a leisurely loop past the Grand Canyon Caverns and the burg of Hackberry before rolling into Kingman. We could follow this path, and, with little traffic, we would have a relatively easy day.

I supported three electric bike riders on this day: Sean, thrilled to be out of the supply truck, Dominique who seemed, by now, tired of the scooter and who floated between numerous vehicles, and, of course, Ben Hopkins.

With a gradual downhill grade and beautiful weather, we made good time. Considering it was August, I had expected it to be blistering hot, but

Scenery as we came out of Seligman, with higher plateaus to our north.

somehow we got lucky. Well, sort of. The high in Seligman on this day was eighty degrees; the high in Kingman was ninety-nine degrees. So, as we went down in elevation, the temperature was going up—almost twenty degrees! Historically, this contrast appears to correlate—i.e., that the high desert runs about eighteen–twenty degrees cooler than the low desert.

Most of our ride on this day was interesting, but uneventful. A lot of rock and a lot of dry desert. Picturesque at times, boring at others. But on we rode.

As we reached Peach Springs, I went ahead into the small town to look for a place to eat. I didn't find much, and, in the course of doing so, I lost track of the bikers. Not to worry, though; I found them back on Route 66 just up the road at Hualapai Lodge, an American Indian hotel and restaurant.

Hualapai Lodge has an interesting American Indian décor and fla-vor while offering attractive overnight amenities. In Peach Springs, we found some good food and nice atmosphere to chitchat at Diamond Creek Restaurant. In such a small town, this place was a real retreat. www.grandcanyonwest.com/lodge.html

After a quick bite, we were back on the road again. Traffic was not a problem—in part because this area was really desolate, with few homes or businesses. In fact, at times there seemed to be more remnants of previous inhabitants than there were of the current ones. But for us, it was unique and interesting territory as we motored through the canyons and valleys at 15 mph.

With the downhill ride, there wasn't much need to swap out batteries,

Descending the desert highlands from Seligman to Kingman.

Dominique, Sean, and BenHop minimize wind resistance as they roll downhill into Peach Springs.

but eventually they did run out. So we found some shade and took a break and swapped all of them out. While doing so, we looked around the terrain and spied someone's large, luxury home all but hidden in the rocks. Who the owner was or why the home was there, we had no idea. But it was there, even if no one else knew.

After our break, we headed on down the road, winding through the

Sean and Dominique enjoy the dusty, arid desert air.

rocks and eventually into the small town of Hackberry. Not much there except an old general store (http://hackberrygeneralstore.com), so we kept on going. After Hackberry, it was a fairly straight shot on into Kingman, where we arrived in the late afternoon.

There was *one* intriguing stop just before we got into town, however, that should be pointed out. As we were riding along just above the city, we came across this huge, green statue. Was it a . . . a giant tiki head? Yep, Giganticus Headicus is a fourteen-foot tall tiki head at Route 66 and Antares Road that was created by Gregg Arnold.

An award-winning artist, Mr. Arnold built the tiki head with metal, wood, chicken wire, Styrofoam, and cement. It sits outside one of his studios. Why did he build it? He's an artist. Enough said.

Taking a break in one of the few areas of shade during the day.

The giant tiki head of GiganticusHeadicus.com. Remember Night at the Museum Dum Dum?
www.facebook.com/gheadicus

Green Lightning *sits below a high ridge of rocks at our rest stop.*

Once we found the Days Inn (our overnight stop) and checked in, I charged the LEAF through the window of my room. We didn't have an event in Kingman, so we all got a chance to relax and do our own thing for dinner and the evening. Tomorrow would be another milestone as a group when we would head into Nevada and Las Vegas.

CHAPTER 35

Day 32: Kingman, AZ, to Las Vegas, NV

"Burgers and Bullets and Bailin'."

"No Permit = No Tape It = Dammit!

"What Happens in Vegas, Sustains in Vegas."

It was Sunday, August 4. Why couldn't our schedule have gotten us to Las Vegas on a Saturday night instead of a Sunday? Sunday was likely to be pretty dull. Nevertheless, it would be a good day as we rode triumphantly into "Sin City" I knew we had an event in Vegas. Maybe we'd get the key to the city, and I'd hit a Jackpot and dance with a goddess in a white and gold dress at Caesar's Palace. Hmmm. More likely, we'd get pulled over for cruising a congested area, and I'd end up in a toga after losing my shirt at the slot machines and whining to an uninterested barkeep about my losses. But you gotta dream.

Our distance today was fairly long, at over 105 miles. We'd be arriving in Vegas from the southeast after having crossed Hoover Dam; the entry was significant, we would later learn, since our event was on the northwest side of the city. This would add a lot of additional miles to our day's journey. In fact, about fifty more! We would be going down another 1,300 feet in elevation, which was good, but you can't go down to a dam without having to climb back up out of a river canyon. So it wouldn't all be downhill.

The bikes and I headed out early, as planned. Benswing and I would have to recharge along the way, and we had identified an RV park along the way to recharge. Our route for the day would again take us away from I-40 and head northwest on U.S. 93, a good divided four-lane highway with little traffic. But as we headed out of town, on the last of Route 66 we'd see for a while, we were again reminded of its history with more Kingman Route 66 stores and another aging steam locomotive. Parts of Route 66 are still alive.

Rachel rode with me today as she continued to heal from her fall. I always enjoyed it when she rode along because she was a very positive person and you could tell she truly cares about other people. She is one of the kindest and most caring people I know. Well educated in journalism and communications, she's from Australia, lived in England and now in Thailand, and held roles with the United Nations and similar positions involving kids' education. Clearly, she's a lady wanting to make a difference for kids, yet, in our conversations, I didn't get the impression that she really knew how she wanted to make such a difference. So our conversation of the day would include her injuries, how others in the Tour were getting along, and what her future held—after all, she had quit her last job to come along on this electric parade.

After I supplied a couple of batteries to BenHop on the bike, Benswing caught up to us. Benswing and I then drove ahead so we could recharge at the RV campground. At about fifty-five miles out from Kingman, we found our target: LAST STOP.

Appropriately named since it was the last stop along U.S. 93 before

Atchison, Topeka, and Santa Fe steam locomotive on display
in the heart of Kingman, Arizona.

Heading down U.S. 93 toward White Hills, Arizona, and the Nevada border.

the Nevada border, LAST STOP was an amalgam of tourist necessity and tourist indulgence. Included in their list of services were

*Grand Canyon and tourist info;

* a bar and café;

* a gift shop;

* an RV park and cottages;

* Powerball and Mega Millions Lottery ticket sales;

* a 50 mm machine gun firing range.

Say, what? Yes, that's right, folks. Here in the middle of nowhere is a place where you can have a world-famous hamburger and then head out to shoot off a few rounds with a large machine gun. Pretend you're taking out the Islamic State (aka ISIS or ISIL) or the latest alien creatures. It's a chance to be Rambo or Mr. T. In fact, you can be as macho as you want to be!

LAST STOP caters to Vegas tourists and buses or helicopters them over to Arizona. It's a long day trip.

And how do you find such a place? That's easy. Just go to http://bulletsandburgers.com

If you visit, you might even want to stay overnight at the RV park or at a cottage—located directly beside the firing range for your convenience!

When Benswing and I arrived, we paid a few bucks for the electric power we would use, and the manager told us to go into the RV park—which was all gravel and open field—and hook up. So we went into the park and started preparing to charge. We weren't there long before a helicopter suddenly flew in and landed about 100 yards from us, bringing in another special customer; at least, I assume it was.

A few minutes after that commotion, the machine gun suddenly burst out loudly about 150 yards from me: RAT-TAT-TAT-TAT-TAT! RAT-A-TAT-TAT-TAT! OMG! I about crapped in my pants! I know the shooter wasn't aiming at me, but the power of that gun and my knowing it was so close and pointing in my "general" direction, with or without barriers between us, was scary as hell. I was ducking most of the time the gun was firing! Needless to say, I wrapped up the charging setup in record time and headed for the grille—i.e. I bailed. I did not want to be around if a bullet got through the barrier or a "stray" came my way.

I cannot imagine staying in an RV overnight there.

LAST STOP caters to Las Vegas visitors.

After I'd gone to bathroom, BenHop arrived, and we all had lunch at the restaurant. The food/burgers were pretty good, I have to admit. A short while later, the rest of the Tour arrived, including Susan and the documentary crew. We planned to meet here so that all of us would be together to cross the Nevada state line together and to film at Hoover Dam.

While the others had drinks, Jonathan, Susan, and I got into a discussion about contacting Nissan again to ask if it would sponsor our Tour and/or the documentary. Since we were only a week away from the West Coast and could claim a cross-continent victory and potential Guinness World Record, it seemed like a

Bullets&Burgers.com provides you with various "Machine Gun" options.

Rachel and I practicing our best machine gunner techniques.

good time to ask. So Jonathan and I crafted an e-mail and sent it to the VP of Marketing on the spot. I thought it was pretty well written, and we hoped Nissan would jump on board even if it was late in the game. It was time for Nissan to "bite the bullet" and to step up to the sponsorship plate, we thought.

While Jonathan and I drafted the e-mail to Nissan, Susan went back to her ongoing messaging-with-future-event-site contacts. But, on this day, one of the conversations wasn't going so well. The contact didn't seem interested in an event. In fact, it didn't know anything about it. Turns out, it was Google.

Say what? Google headquarters was our final destination. From Day 1, Susan had told us that it's where we would end our trip and that there would be a big celebration because Google supported our efforts to use alternative fuels. It is a progressive, forward-looking company and was rumored to be developing its own electric car. We discussed giving Google "our" car if we completed our trip and could talk Nissan into it. It would go with the energy-efficient Google community that Google had in Mountain View, California. How could Google not know about our plans?

Well, it turns out that Susan hadn't really contacted Google yet, and it turned out that it had a big electric-vehicle event it was hosting the weekend after our planned arrival. Google didn't want to greet us!

This was a serious blow, both physically and mentally. All along we had a goal. A target to achieve. But now? Where would we go? Who was going to greet us with open arms and laud our accomplishment? Arrggh.

Susan assured us she and Tiffany would work something out. Our PR super helper Tiffany was on it, and she would talk Google into letting us finish our trip at its headquarters. We left it in Susan's and Tiffany's hands and pressed on.

We'd spent too much time here, and we'd charged the LEAF and the motorcycle enough for the day, so we packed up to head on down the highway. A couple of people from the Tour did fire the machine gun for fun; it wasn't something I needed to do. Maybe I'm a pacifist. Maybe ducking wasn't my idea of fun.

It was only another thirty miles to Hoover Dam, and the miles went by quickly. There was a rest stop and overlook just before we descended to the dam area, so we stopped and took a couple of pictures. But soon BenHop was passing by us, so we hustled to catch up to him before he reached the dam.

The road down to the dam was pretty steep and winding; traffic was considerable since this was the height of tourist season. We got down to the top of the dam and slowly crossed over; there were plenty of pedestrians and congestion, so it was a slow procession. When we got back to the Arizona side, we found parking, and we all hiked down to dam level.

So here we were, taking in the view as we slowly walked across Hoover Dam with the videocamera rolling. We were ready to celebrate crossing the state line into Nevada when were promptly interrupted by the police again, more specifically, by the U.S. Bureau of Reclamation. The officials informed Susan and Jonathan that they could not shoot video on the dam

LAST STOP is a great "Gas and Blow" for the Rambos of the world.
www.facebook.com/arizonalaststop

*A U-nique ad for gas; if you want to take out a few
aliens today, they have the gun for you!*

*Super character images are plentiful at
Bullets&Burgers.com, and Rachel likes cool
characters.*

without a proper permit, which had to be applied for in advance. It was a law—enforced, in part, for the Department of Homeland Security, I think. But this wasn't new to the Tour; however, paying for a permit was expensive, so we didn't pay for one. We'd gotten away with shooting video at the Grand Canyon. This time, we got busted for it.

Well, so much for adding to our video library. Evan and George had

<hr />

Almost exactly a year after our visit, a thirty-nine-year-old instructor was accidentally shot and killed at LAST STOP by a nine-year-old girl being trained on the use of an Uzi.

A beautiful view near the Arizona-Nevada border,
where Rachel takes my picture.

Scenic views of the desert and the Colorado River
as we approach Hoover Dam.

to return their equipment to the van. The rest of us continued across the dam, pausing to take lots of pictures and taking in the great views. We all enjoyed staring down at the height of the dam. Hoover is such an immense structure it kind of freaks you out the first time you see it. We took our

Rachel poses for a pic before we head to Hoover Dam, and a shot of the Colorado River.

A new bridge on U.S. 93 (ahead) caused us confusion in getting to the dam.

Crossing the Hoover Dam back to Arizona.

The group hikes down to the dam and finds time for fun as we cross.

time absorbing the scene: the blue lake, the tan-colored, enormous dam, the multicolored rock hills, etc.

Formerly known as Boulder Dam, the Hoover Dam was built during the Great Depression from 1931 to 1936. A concrete arch-gravity dam, it stretches across the Black Canyon between Arizona and Nevada and holds back the Colorado River. Building it was a massive effort, involving thousands of workers and their families; over one hundred of the workers actually lost their lives. It was the largest dam in the world at that time.

The dam controls floods, provides irrigation water, and produces hydroelectric power for public and private utilities in Nevada, Arizona, and California. Hoover Dam impounds Lake Mead, the largest reservoir in the United States (by volume) and was constructed for $49 million. At a then-height of 726 feet and a then-length of 1,244 feet, it has become a major tourist attraction, with nearly a million people touring the dam each year.

A police officer explains to Susan what she needs to be able to shoot video on the dam.

We continued across the dam and eventually made it to the cool air in the gift/snack

241

shop on the Nevada side. It was a crowded weekend for visitors, so after a quick drink, we headed back into the sun. We noted a couple of statues erected on the main level to commemorate the efforts of the skilled climbers who were instrumental in the construction of the dam.

We needed to keep moving to make our evening event in Vegas. So we gathered up on the Arizona side and began our little electric parade to cross the dam and then climb out of the Hoover Dam canyon area.

I'm not sure if Jonathan shot any video during our drive across the dam, but I somehow suspect he did. Traffic wasn't too bad this time, and we easily made our way across, pumping our fists again at the Nevada state line and taking in the sights.

We had to go slow as we climbed out of the canyon since BenHop could only pedal about 15 mph going uphill on the bike. Ben was a strong guy who could pedal up mountains, but this was a steep incline. He did the best he could, but, at least with the electric-power assist, he wasn't stressed doing it!

With our slow speed uphill and since it was a windy two-lane road up to U.S. 93, we did block traffic behind us, and some of drivers did get impatient. There wasn't much we could do, however. Unless there was a place to pull off, there was nowhere for us to stop along the road. We did what we could when we could to let traffic pass. We had learned during our trip that there were going to be impatient drivers. You just had to be courteous when you could and hope some impatient idiot didn't cause an accident and blame you for it.

The Ride the Future Tour pauses to enjoy the view from Hoover Dam.

The immenseness of Hoover Dam; those are cars and people at the top. www.usbr.gov/lc/ hooverdam

View upriver of Black Canyon and the Colorado River.
https://en.wikipedia.org/wiki/Hoover_Dam

The new highway bypass crosses over one of the mammoth water drains.

Once we reached U.S. 93, we headed west into Boulder City and toward Vegas, which was only about thirty-five miles away. We stayed on U.S. 93 as it wound up around the hills and intersected with another highway (U.S. 95), and we headed into Henderson, a sizeable city just outside of Vegas. At this point, traffic had picked up considerably and U.S. 93 became an interstate. We wanted to avoid this and immediately exited to Route 582, which was a straight-line shot into Las Vegas. That was the good news; the not-so-good news was that the route held only city traffic, involving a lot of starts and stops for traffic lights, pedestrians, and business traffic.

The really bad news was that we were approaching the city from the far east side and that our event was on the very far west side and that we were hitting all this city traffic in late afternoon at rush hour. Ugh. On the interstate, the ride would have been easy. But in city traffic and at 105 degrees, it was brutal for everyone—even me.

After what seemed like forever, we finally made it to the event. We were late, however, and the press was waiting for us. The location was Las Vegas Cyclery which specializes in everything bikes! From repair and parts to clothing and all types of bicycles to organized guided tours, Las Vegas Cyclery is your source. It even rents bikes and sponsors a local cycling team.

In addition to being bike experts, however, LVC is also very big into giving back. Jared and Heather Fisher, the owners, built the business from ground up, and it thrived. So much so that when the business hit its twenty-year anniversary of success, the owners wanted to give back in the best way they could think of to make a positive impact to their community, the environment, their friends, and their coworkers.

The result: they built the first Net Zero Energy bike shop in the world!

With a goal to strive for the highest environmental rating possible, LEED Platinum, they built a beautiful facility that not only sells beautiful bicycles but that also sells them without requiring any resources from the outside world.

The Tour met the waiting press, answered their questions, and showed them our vehicles, as well as explained our purpose. We met some of the employees and customers of LVC, who were very enthusiastic supporters of being environmentally conscious and sustainability-minded. And after the press got their stories, the owners invited us to join them in their upstairs conference room, where we all sat in a large circle. The Tour talked about its adventures and our members' backgrounds, while the LVC folks talked about their facility and activities in sustainability. It was a very casual and very interesting discussion.

For example, the LVC building is not your typical construction. It was made with recycled and locally produced materials (within 500 miles max). It is completely powered by the sun and wind (with a 50kW solar PV system and a 5kW wind generator). It was designed to provide 100 percent of its energy needs throughout the calendar year; this is called a "Net Zero Energy Building."

Essentially, the LVC building is its own power plant and can sell excess power to the utility company. In addition, the owners and their employees are big on the education of others regarding sustainability, thus taking a leadership role. In fact, they have a written tour of the building, which will no doubt give you a better insight into sustainability; check this out!

http://lasvegascyclery.com/about/green-energy-
building-tour-pg132.htm

*Finishing our drive across Hoover Dam at a
leisurely pace and climbing up to U.S. 93.*

My "trick" shots of the Tour, the traffic, and the scenery behind me as we wind our way up to U.S. 93.

When we finally wrapped things up at LVC and said our good-byes, it was starting to get dark. Luckily, the hotel was nearby, and we quickly unpacked our gear. But we couldn't make a trip to Vegas without *doing* Vegas! So we all got back in the saddle and headed back to downtown—over ten miles away! Not hard for me, but, for BenHop, it was a big commitment—of over twenty miles. But he did it.

So, for the next one and a-half hours, we drove to the "new" Vegas strip and did our electric parade through the main streets of Vegas, hollering, honking, and absorbing the night life of Vegas. We certainly got looks and interest, but mainly it was just fun.

Here, we didn't need no stinkin' permit for Vegas, so Evan filmed us the whole time. He specializes in taking shots from various angles and isn't shy about sitting outside a car door for a shot or dropping the camera to street level if he thought the shot would look good. And actually, that is *exactly* what makes him good at what he does.

We finally grew weary of cruising, and a couple of us stopped to get a bite to eat at a casino; the rest went on to gamble or to party. As it was, we ate between 10 and 11 p.m., and I was tired. It had been a long day.

When we finished our meal, a small group of us headed back to the hotel, the Element Las Vegas Summerlin, another very eco-conscious hotel. We finally arrived around midnight, having driven/ridden 155 miles for the day. I located the ChargePoint charging station at the hotel and plugged in overnight.

www.starwoodhotels.com/element/experience/green_vision.html

As LVC showed us, "What happens in Vegas, stays in Vegas," and also, "Sustains in Vegas." But I could not sustain. I didn't party. I didn't gamble. I didn't watch a show. I went to bed.

Mingling with the press and meeting environmentally conscious LVC customers in Las Vegas.

Employees, friends, and customers of LVC share a photo moment with us. http://lasvegascyclery.com

Left: Benswing playing at the LVC windmill. Amid stretch limos and pickup trucks, we parade through the heart of Vegas, baby!

CHAPTER 36

Day 33: Las Vegas, NV, to Baker, CA

"Taking Recharging to New Heights."

"Cut Off and Kicked Off."

"Pool Party Pestilence."

Ben Hopkins was a pain in the ass. Yes, when he wasn't the simple, funny bear of a man that people love, he was a pain. Not that he intended to be. He was never rude to anyone or disrespectful in any way. With his big frame and strong arms and legs, during the Tour, he nearly became the image of an unkempt, unshaven caveman, which was somehow endearing. But that wasn't why he was a pain; he was a pain because Ben was not the most organized guy in the world—far from it. And at times, like today, it cost him.

After a long day and a late night yesterday, came another sizeable leg today of ninety-five miles from Las Vegas to Baker, California. To enhance our adventure would be a temperature of nearly 105 degrees. Even though we were all tired, it was important that we leave Vegas early to avoid as much afternoon heat as we could. BenHop was one of the early-to-bed folks like me and was ready to leave the hotel around 9:00 a.m. We discussed the route, as we always did before he left; I usually gave him a sizeable lead before I tracked him down and followed him with a replacement battery. We agreed on the route, and he soon took off by himself.

The route today was another interstate challenge. If we could have taken the bypass around the city, we could have easily hooked up with our highway south and west. But since we could not take the bypass, we had to take main roads first east and then south as far as we could until we *had* to take our chances on the interstate, which was I-15. I say "had to" since our only other option besides a freeway would have added thirty-five miles to the route through open desert—a significant risk to the riders and a lot of extra work for Ben on the bike.

After Ben headed out, I spent time at the hotel having coffee and talking to Rachel and others that were up and about. Now that we were heading into California, I knew that most of my recharging needs could be met with public chargers along the way, especially when we reached the coast. I had only a couple of unknowns left, Baker, our stop for tonight, and one other huge "hole" north of San Luis Obispo since the Tour wanted to take the scenic Pacific Coast Highway (California SR 1) along the ocean into Carmel and San Francisco. With Susan's help, I had a lead on that "hole"— my biggest concern for the entire trip.

Forty-five minutes after Ben headed out, I got a phone call; it was Ben. He had only gotten a couple of miles away when his battery died on the bike—probably it never got charged properly or was mis-marked as charged. (We used little colored stars to indicate charged batteries, but occasionally one might get a star it didn't deserve.) However, Ben had been unable to call me because he didn't have a cell phone. And since he was at a busy intersection, he couldn't easily borrow one from someone to call me for a replacement battery. So why didn't Ben have a cell phone? Well, because he's Ben. He had been given a throwaway phone by his sponsoring bike company, but it had run out of minutes.

Rather than ask for another or buy another, he took his chances and simply didn't do so. And, on occasion, his habit became problematic. Like when he got thrown off the interstate by the highway patrol and wasn't where I thought he would be. After losing him for an hour or more, he finally borrowed a phone and called me so I could bring him a new battery. Today, was a similar day.

I grabbed charged batteries from Sean in the supply truck and rushed over to Ben. I found him at a major busy intersection standing with his bike under a small tree that offered the only bit of shade around. But the tree happened to be one with thousands of small thorns in it, many of which were on the ground, right where Ben had parked the bike to wait for me. He soon found that he not only needed a battery; he also needed tools and the tire pump because he had two holes in the rear tire. So much for an early start.

In the morning heat, Ben tore the bike apart, located the holes and thorns in the inner tube, and repaired them. I didn't help beyond tool support, but I was hot just watching. I felt bad that I hadn't been right there for an immediate battery swap, which could have been made on the street. But I couldn't be with him all the time, and Ben's not having a phone cost him extra work and time.

When Ben had finished his repairs, we started off together via major roads in the city. We made our way to South Las Vegas Boulevard—the old Vegas strip—and took it all the way to the edge of the city. And since it was a busy street, I had to play leapfrog with Ben and would go ahead of him and wait for him to come by. Near the edge of the city, I staked out a large parking lot and began listening to music while half dozing off. I waited for Ben, and waited, and waited, but he didn't come by. Finally, I went looking for him. I was pretty sure he hadn't passed me. Eventually I found him, pedaling hard. Once again, the battery had died, and he didn't have a way to reach me. We swapped out the battery, and he was off again. Not the most efficient process, but we were still moving.

We continued following South Las Vegas Boulevard as it went out of town, and we continued through the small town of Jean and eventually to the town of Primm and Whiskey Pete's Hotel & Casino, an icon establishment on the California/Nevada border. Now some forty-six miles from our Vegas hotel, Primm would be our spot for lunch and to recharge the LEAF. Benswing also met us here so we could recharge together.

When we arrived, BenHop went looking for a drink, while Benswing and I went looking for the town's recharging stations. The town had two, so we could charge simultaneously. Or so we thought. It turned out that, as we had found occasionally along our trip, one of them was not working. Mine. And since Ben was already hooked up and charging, we started looking for some way for me to charge also while we took a break. We looked all around, but couldn't find an outside outlet anywhere. Finally, we went into a parking garage and found one. But not where we had expected it. Instead of being along the base or middle of the wall, these outlets were in the ceiling! And the ceiling was about twelve feet high.

Finding the outlet was one thing; connecting to it was something different. Finally, I helped Benswing (who's probably six feet four inches tall) climb up on the A-pillar of the LEAF, making sure not to step on the roof—which undoubtedly would have been dented. With my help, he was able to stretch and reach up to plug in my charge cord for 110v. It wasn't the best, but some power was better than none, and I could finish on the 220v public charger Ben was using after he was done.

The two Bens and I had lunch then at Whiskey Pete's Hotel & Casino. As much as anything, we just appreciated the air conditioning; it was hot outside now. While eating, Benswing mentioned the large solar panel farm that we had seen far away as we rode in. This farm is a part of First Solar Electric, a large solar power equipment supplier, and the farm is probably the largest setup of solar panels that I've ever seen. Appropriately placed in the desert, where there's lots of sun, I would expect it could provide a lot of power to Nevada and California.

We decided to try to get a better look at it, so, after lunch, we climbed to the top of the parking garage—the highest structure around—to see what we could. From the top, we could see all the panels glistening in the sun and a bright beacon of light from a tower. I'm not sure what the light was for, other than attention, but it was clearly visible. Regardless, it seemed like a great investment and location for an alternative power source to meet society's needs.

The Bens and I returned to the parking garage, disconnected the LEAF from the ceiling, and I drove the LEAF over to the public charger. Benswing's motorcycle was sufficiently charged to make it to Baker, so we started charging mine for a while. Benswing tried

Finding an outlet at Whiskey Pete's Hotel & Casino was one thing; connecting to it was a whole different challenge.
www.primmvalleyresorts.com

251

to charge some more on 110v by taking it inside the casino, but the security guard threw him out! Though Benswing's motorcycle was electric and had no emissions, taking it inside was against the rules. Hmmm. I wonder if he would have thrown out a vehicle from his own era, like maybe Fred Flintstone's car? I digress.

We sat twiddling our thumbs and chatting for perhaps an hour and decided we should be able to make it to Baker. Coming out of Vegas at 2,000 feet, we'd climbed almost 700 feet on a long, steady climb. We had a way to go before we would go over the mountain pass, at around 3,000 feet, and would drop back down to Baker, at only 1,000 feet. The last part would be easy.

We had expected to meet up with the scooteristas at this point, but that didn't happen. I'm not sure if it was too much late-night Vegas or what, but they were behind us. Jonathan and the documentary crew did show up, however, and Jonathan rode with me to Baker so he could film some of me driving and talking about some of the LEAF's features like regenerative braking and its acceleration and excellent torque.

Even without the scooteristas, we decided to press on. We still had a long way to go and would now be taking a risk. BenHop's riding along the interstate was illegal because he rode too slowly to keep up with traffic. That meant that neither I nor Benswing would be able to stay with

*Close-up of First Solar's farm, with power lines
running to nearby Higgins Generation Station.*
www.firstsolar.com

The Bens and I talk charging challenges and the road ahead at Whiskey Pete's Hotel & Casino.

him; we'd have to play leapfrog. The further challenge was that there were no services of any kind for our remaining fifty plus miles, so if anything happened, we had a long way to go to get help or to buy what we needed. Worse yet, if BenHop got caught by the California Highway Patrol, we had no idea what would happen to him. There was no nearby road for him to use.

We got onto the interstate on ramp, did a few more fist pumps as we passed the Welcome to California sign, and started up the long, hard mountain grade. Here, unlike before, I had to do normal traffic speed, which on I-15 was a minimum of 70 mph, or you got run over. This was problematic because the LEAF used a lot more power at that speed. With us going up a steep hill, the LEAF's power was drained quickly; I could watch the "power bars" on the dash of the LEAF disappear as I drove. I did pull over to wait for BenHop to catch up along the way, and, at one point, he needed a battery and had to pedal uphill a short ways to me.

We kept leapfrogging until finally we reached the top of the mountain pass. By this time, my battery had lost a lot of power. In fact, as Jonathan videotaped, I told him that my car said I had enough battery power to go another twenty-two miles but that the odometer said I had another thirty-three miles to go! Now I wished I'd gotten a full charge back at Whiskey Pete's, but it was too late now. All I could hope was that going downhill with Nissan's regenerative braking feature would leave me enough power to make it all the way to our stop in Baker. Fortunately, it did!

We flew down the back side of the pass. It was steep, and it was fast. But as BenHop would later put it, it was "like dropping down into a furnace"

BenHop makes the long, hard climb out of Vegas and Primm to the top of the mountain pass.

from the pass—it was a lot hotter at the bottom. About two miles outside Baker, we could finally get off the interstate, and we took the straight road right into downtown. As we neared downtown, I hollered over to Ben that we were coming up on the World's Tallest Thermometer, which he found interesting. Less interesting was the temperature itself. I think it was like 104 degrees. The thermometer is a tourist attraction of sorts. It stands 134 feet tall, weighs 76,000 pounds, and contains 125 cubic yards of concrete! But for us today, it was too hot to warrant a visit. We headed to the motel, and, hopefully, a pool!

www.worldstallestthermometer.com

When we arrived in downtown Baker, we had beaten the scooter riders there, so we had to wait for Susan regarding our rooms at the motel. The problem was that no one knew where Susan was. Cell phone service was not good outside of Vegas and in the California desert; it was almost like being back in the New Mexico and Arizona. Fortunately, not too much later, here came Susan and Dominique. It turned out that they had gotten on the Vegas bypass and were immediately stopped by the highway patrol, who threw them off the freeway. They ended up taking the long way round on the back roads in the middle of the desert, at over 100 degrees—some thirty-five extra miles of riding—and all by themselves. Thank God, they didn't get lost or have a problem. I don't know when we would have found them. Regardless, they had no Whiskey Pete's break from the heat, so the first thing they did when they arrived was run and jump directly into the swimming pool—clothes and all!

Susan got us into the motel; the Wills Fargo Motel. At first blush, this motel looked like an interesting and historic place to stay. Did stagecoaches stop here as they crossed the desert to Los Angeles from the east? Well, I believe there was some historic significance to the location, but "interesting" is not the word I would use to describe our accommodations. There was a bed, I'll give it that.

The first thing I did was look for a way to charge the LEAF. There was no easy way to charge from our room, so the owners told us to use the 220v outlet in the laundry building where the dryer was. And so we did, and I had a cord to fit it. So the LEAF got "laundered" on this night.

So maybe the rooms weren't the best, but we could have a great pool party of our own, right? We ordered pizza and food and had it delivered poolside, and everyone jumped in! Sean brought out a boom box, and we enjoyed music and the cool water; we had a great time. There was only one downside. We had to share our pool with other guests. Cockroaches. Not the big ones, but young ones. Hundreds of them. They were everywhere around the pool that evening. Just keeping our food away from them was a challenge.

Nevertheless, we made the best of it and enjoyed our party. It was another stress-relief event we needed. That said, I shudder every time I think of what my room must have looked like that night after I turned off the light. Eeek!

Green Lightning gets clean energy from the
laundry room at Wills Fargo Motel.
www.facebook.com/pages/Wills-Fargo-Motel

CHAPTER 37

Day 34: Baker to Barstow

"Peggy Sue, We Love You!"
"EZ + A/C = 0"

Our next stop was Barstow, a small desert city with a few outlet malls because people from LA like to have a place to shop on their way to Vegas. But Baker and Barstow were not originally part of the Tour route. Susan had wanted to drop due south out of Las Vegas and ride through the middle of the Mojave Desert Preserve and into Twenty-Nine Palms. While the trip was shorter in distance, when I looked into it in depth, I found there were no sizeable towns at all along the way and no services of any type that I could see in the seventy-three miles we would ride through the Mojave Preserve itself. And doing it in the middle of August? That seemed like a recipe for disaster to me. The safer play in my book was to go west through Baker and Barstow and then drop south through Big Bear to Orange County and LA. After some considerable debate, Susan finally agreed, and today we would take the safer route.

It was less than seventy-five miles to Barstow, but we would have to ride the interstate once again for the first twenty-eight of it. So far, we'd been lucky and BenHop hadn't been stopped by the highway patrol. Today we'd continue to leapfrog until we could exit and take Yermo Road into Yermo and all the way to Barstow.

Rachel was riding again with me today, and we left shortly after BenHop did, which was around 8:00 a.m. It would be a very hot 105 degrees in the desert, and since it was a short-distance day, I decided to splurge with Rachel: we'd turn on the air conditioning! Yippee! This was a rare luxury since the AC used a considerable amount of power. And since I always try to preserve my power for "just in case" situations, I never used it. But today just seemed like a day we could enjoy it without worry.

Around an hour and-half later, Ben slowed because he needed a battery. We gave him a fresh one and took a few minutes to enjoy the desert. It was comfortable since it was still morning, and the area was actually picturesque. Unlike the New Mexico and Arizona deserts, which are at higher elevations and consist of dry earth and rocks, the Mojave Desert is a lot of sand from mountain erosion and a former gigantic sea bed. But water was scarce in all the deserts.

After a water break, a short rest, and a couple of pictures at an exit ramp, we got going again. I gave Ben a big lead and then eventually pulled back onto the interstate. Rachel and I continued with our conversations in comfort.

We finally got to the exit where we could ride parallel to the interstate. This was a relief.

Rachel and I enjoying our drive in the desert.

However, I now had a new concern. Barstow was located at a higher elevation than Baker, some 1,200 feet higher, which made the interstate grade pretty steep. This, coupled with my high highway speed *and* my air conditioning over a sustained time, had drained my battery at a rather alarming rate. So, about halfway to Yermo, I shut off the AC.

Meanwhile, we made good time on a road with not much traffic. Soon we came across a huge parking lot of U.S. taxpayer dollars: a large army supply and storage area. Here were hundreds of Hummers along with a lot of other equipment parked in the desert. Sitting there idle, they looked like they were waiting for a war.

A little further down the road in Yermo itself, we found Peggy Sue's 50's Diner. We decided to try it for lunch and BenHop, Rachel, and I pulled in. Peggy Sue's is a real throwback to the diners of the 1950s, complete with

*Taking a break in the desert with my wetted towel to keep cool,
as Rachel plays in the sand.*

A huge army storage base greeted us along the road to Yermo.

the waitress costumes and 1950s memorabilia. They also have a five-and-dime store and specialize in old-fashioned ice cream shakes and malts. Built in 1954, it was restored in 1981 by the current owners, with three booths and nine counter stools. Just like the old days.

When we'd had our fill of the waitresses (they were a lot of fun!) and

Peggy Sue's 50's Diner has good food and a great atmosphere.

finished our lunch, we headed out again along Yermo Road. We didn't get far, however, before we ran into an obstacle. It was a gated outpost. Thinking perhaps we just had to drive through it since there were no warning signs that the road ended, I drove up to the gate and asked. I was promptly told that I couldn't drive through it. We turned around and headed for the interstate again, which was our only other alternative. There was a stretch of about five miles where there was no side road that Ben would have to go through without being seen. However, the motel we were booked was on the far side of Barstow, and the shortest way to get there by far was to stay on the interstate, and Ben decided to just do that.

I was going to go ahead and try to make it to the motel since there were no public chargers in Barstow (probably because the folks in charge there never expected anyone with a LEAF to ever drive through the city). More interstate driving, however, meant faster drain on the battery again, and I was growing very concerned; I was on the last bar of power. I knew it was my own fault for using the AC, but, at this point, I just wanted to get to a place with 220v.

Rachel and I left Ben and headed to the motel. I made it to the edge of Barstow when my last bar went out. I was in new territory now. I knew I could go into turtle mode at any minute. I decided to chance it and stay on the interstate because it was the most direct route. We held our breath. The number of miles I could go with remaining power went dark. I saw our exit and slowly coaxed the car through a couple of intersections and side streets, and pulled into the motel parking lot. I guess there was something left in the battery, but the gauges didn't show it. I checked in and took the car to the room, where I plugged in overnight. I knew I'd just gotten very lucky. I realized that the assumption of an easy day plus the unknown was risky; i.e. EZ + A/C + ? = 0 power! I knew I didn't want to be that close to running out of power again. So, like the "Soup Natzi" in the famous Seinfeld episode, I decided: "No more AC for *you*!" I don't think I used the AC the rest of the trip.

Meanwhile, BenHop wasn't so lucky. He'd started on the interstate, but Barstow has a state highway patrol barracks, and the patrol officers caught him and threw him off the freeway. So Ben had to take the long way around the city of Barstow to get to the motel, adding probably seven miles and an extra hour to his day.

The others arrived later, and, with no planned event this evening, it was a pretty casual night. I worked on some logistics to meet people in Orange County when we arrived and some other issues. Tomorrow was another potential challenge to plan for. To avoid a very long stretch of interstate into Los Angeles, we would follow a back road that would take us into the San Bernadino Mountains and an overnight stay at Big Bear Lake. And as I had just experienced, uphill grades were a huge drain on the LEAF's power supply. Ugh.

Driving into Barstow was supposed to be easy,
but it turned out to be too close for comfort!

CHAPTER 38

Day 35: Barstow to Big Bear Lake

"Rope-a-Dope Up The Slope."

"Water Stealing Leaves Me Steaming."

After two weeks of riding through the desert, today we would *finally* get out of it. But while that sounded great, climbing mountains was a new environment that we had yet to test the LEAF in. We had done a little bit of it in New Mexico to climb to the upper plateaus of Las Vegas and Santa Fe, but this was different. While our distance was only about sixty-five miles in total, we would be climbing about 4,600 feet from Barstow to Big Bear Lake, with 3,800 feet of it in an eight-mile stretch from just south of Lucerne Valley to the top of the first ridge near a small town called Doble. I was not concerned about the LEAF's ability to climb the steep road. After all, electric vehicles have excellent torque capability, but how fast would I deplete battery power doing so? That, I didn't know.

Several of the team were up early in the morning, and I found BenHop in the motel lobby. I had a little of the continental breakfast and was expecting Ben to be leaving shortly, after which I would follow. But Ben was distracted this morning, and a bit frustrated. His bike sponsor and the PR folks were pressing him to do Facebook and Twitter posts since there were people who wanted to follow his adventure and success. But Ben wasn't the most savvy guy around a computer. A couple of people, including Rachel, were helping him do some posts. From my perspective, the posts weren't pretty. But I gave Ben credit for trying, and I think eventually he got the hang of it.

Meanwhile, I decided to head over to a nearby truck wash for eighteen-wheelers (big in Barstow stopovers) and wash *Green Lightning*. It was time. I used the self-wash and the vacuum, and the LEAF was gleaming again. I headed back to the motel, and, by this time, BenHop was ready. With Rachel riding with me, we took off.

To avoid the interstate, we had to take some back roads. There were always question marks on these roads' conditions, and, in fact, the first one was a dirt road which, we opted not to take. The second one past the outlet stores, however, was paved, at least, and we found our way over to California SR 247 (Barstow Road) and headed almost due south toward the city of Lucerne Valley.

California SR 247 is a long, slow grade of some 800 feet from Barstow to Lucerne Valley. The ground is flat, and there's not much of anything around—just desert, which probably explains why, when BenHop had a flat tire about halfway to Lucerne Valley, we were in the mood to have a little fun with the camera.

Ben needed some tools to make his repair, which I didn't have, so I called Sean, and he rushed over with the supply truck. While Ben worked away, Sean, Rachel, and I started taking some fun pictures, just to pass the time. And when Ben finished, he got in on it. Nothing crazy. Just fun. There was almost no traffic on this road, so we had it all to ourselves. It was an opportunity to grab some "King of the Road" macho pictures!

Back on the road again, we made good time into Lucerne Valley, a small town at the base of the San Bernardino Mountains. We actually had

Ground-level view of California SR 247 looking
back toward Barstow—long and straight!

The look south toward Lucerne Valley and the
mountains at Big Bear Lake.

to descend into the town, which was beneficial since I got some power back that I had used during the climb. We decided to stop at a small café at the intersection of California SR 247 and California SR 18, called Café 247.

The sign on the window of Café 247 reads Greatest Hamburger on Earth, and, for a place in the middle of nowhere, it sure was busy. We found out why. The café had tasty sandwiches, and we had a very good lunch. It had a mom-and-pop atmosphere, with plain tables and simple

'You're being truck-jacked, son. Make a move and I'll fill your cab full of water!'

Armed and Dangerous: ready to take on . . . a nest of fire ants! Call me: The Ex-terminator!

Ben Hopkins: King of the
Electric Biking Road.

Looking back down toward Barstow as
we approach Lucerne Valley.

Overview of Café 247 (courtesy of its Face Book page)
and its shirt promoting its location.
www.facebook.com/Cafe247LucerneValley

signs for messages like The Daily Special, but who cares, as long as the food is good? And since the café sits in the middle of nowhere, why not promote it as such?

With a decent rest, it was now time to take on the mountains. From Barstow, they didn't look so bad. But now, from the base, they looked pretty damn big! After a short jog from the intersection where we had picked up California SR 18, we started our ascent. As we started uphill, I tried to stay closer to BenHop. Even with his bike's electric power, I knew the ascent would be a challenge for him. He was strong and could be an animal on that bike, but these mountains would still be tough.

After about eight miles, we reached the San Bernardino National Forest and the base of the mountains. We'd now traveled about forty-two miles, which was two-thirds of the way to Big Bear Lake; so, with one-third of our miles to go and my battery indicating between one-half and three-fourths of my power remaining, I should be okay, I hoped.

We started our climb, and it wasn't long before we were in the middle of steep 6 percent–10 percent grades, switchbacks, and a very narrow two-lane road. This was not your average gradual mountain climb; this was a cruel "climb me if you can" road. I later learned that this road is one of the steepest highways in the country.

Whether it is or isn't one of the steepest highways, it took its toll. At points, even with a fresh battery, the A2B bike motor simply couldn't overcome the grade, so Ben was forced to power up with his own leg power. It was slow and arduous as he weaved his way back and forth across the road to maintain momentum in between thrusts to move further uphill. And since the road was narrow and there was a little traffic, I couldn't always stay right with him, sometimes driving ahead of him and waiting for him at a pullout.

To his credit, Ben didn't give up. He powered his way up—foot by foot, yard by yard, and mile by mile. He didn't need the encouragement we yelled; when Ben was focused, he could take on any terrain; he blanked everything else out.

While Ben was gutting it out, two questions had come to my mind:

First, where was the documentary crew? This was arguably one of the best examples of the hardship Ben (and the LEAF) had to go through in our quest for a Guinness World Record, and the crew was nowhere to be found. Why?

Its absence was starting to gnaw away at me. Jonathan and the crew enjoyed finding other activities to film and do while we were en route. Often Ben Rich and Susan or others would join in. They had done this several places along the way. But since I was supporting BenHop on the bike and traveling much slower, I could never join them. And the crew's absence also meant that Jonathan spent a lot more time filming the other members of the Tour. No big deal, but BenHop's arduous ascent up California SR

BenHop makes the initial ascent out of Lucerne Valley, which wasn't steep but was steady.

18, as well as the LEAF's, was a missed opportunity for the documentary, as I saw it, and it bothered me.

The second question was about my power. The steep grades were chewing up my battery life rapidly. What once looked good didn't look good at all anymore. I was growing more worried by the minute; the thought of running out of power on the side of the mountain road entered my mind. I knew it was going to be close.

We took short breaks to give Ben a rest, but they weren't long. Rachel and I spent time encouraging and filming him, as well as talking about Tour activities.

Up, up, and up we went. Ben had to be getting tired, and I was down to my last bar of power. We'd done over six miles of this steep-grade climb, and we still weren't to the top. The road was now curving around the mountain, so there was a little relief, but it was still going up!

We stopped again at what we thought was perhaps the top, because the road appeared to be turning west ahead of us. This gave us one last look down. It was a magnificent view. We were so glad that this climb was almost over—one way or another. Ben was proud and in a good mood, as he should have been for his accomplishment. But we still weren't to the top yet!

We pushed on again, and we followed the road uphill as it turned to

BenHop grinds away on the steep grades of California SR 18 to Big Bear Lake.

Fooling around on the steep incline, I'm yelling, "Forward, ho, we will go!"

the west, a big long sweeping curve. I was out of power bars now, and all I could do now was hope we got to the top before the power ran out. If not, I'd have to call Sean with the supply truck and half unload the whole truck to dig out the gas generator that was in the very front, buried by all kinds of Tour equipment, parts, camping gear, and more. That would not be fun.

For the second day in a row, I'd done what I didn't want to do: run out of power. The only excuse this time was running out of power was less avoidable than using the AC. I didn't know how much power the steep

Ben reaches a temporary respite from the climb.

grades would take. And it was evident now that it was considerable. If I made it to the top today, it was for one reason only—because my speed had been very slow. Had I tried to do normal highway speeds from Barstow, I'd be sitting on the side of the mountain, waiting for Sean.

We came around another uphill grade curve, and a sign appeared to our right—a sign telling us Doble was only a few miles away. Yes! That meant the apex couldn't be much further. *Maybe* I would make it, after all. Rachel and I got excited. We crossed our fingers as Ben picked up some speed, with the road leveling out a little. And then suddenly,

*Rachel relaxes and looks over where
we've come from on the road below.*

A great view from near the top of our climb.

*Ben and Rachel enjoy the sun and the view as we
approach the top of our climb.*

My bright idea of taking a picture simulating Ben hauling me up the mountain. It didn't work.

we crossed the Pacific Crest Trail, and we were there. We started going *downhill*! Now I'd pick up power from regeneration. Yeehaw! And, with a couple of more dips in the road, it was evident that we'd made it!

I'd taken the LEAF to as extreme a situation as it could handle. And I had pulled it off. I doubt anyone else has even tried to make a 4,600 feet all-day climb with an all-electric car before, but I could now say that I did! It was proof that the LEAF and electric vehicles *could* go anywhere with a little planning and an infrastructure to support them. I didn't have much of an infrastructure to use, but it's possible for the United States to build one. It was a proud moment for me—and for Ben! One of the highlights of the trip. A great feeling of victory I got to share with only Rachel and Ben.

Now we were going downhill, and I was picking up power—at a good clip. Ben was letting the motor carry him as we wound downhill past large pines and greenery. It was no longer desert climate here. I looked at my watch. It was about 2:00 p.m. If we hurried, I could make it to Big Bear Lake. Jonathan or Susan had set up appointments for us to try flyboarding—the new water activity where you flew into the air and then into the lake on a board. It was very new, but Big Bear Lake had it, and a few openings were available, the last at 2:30 p.m. I drove ahead of Ben and pulled into the pier parking lot at about 2:15 p.m. I saw the action on the lake, where Benswing and Sean were trying their luck at it. Jonathan asked me if I wanted to try it; I said, "Yes," and hurried to the car to change.

When I came back, however, Jonathan was being towed out into the lake by the jet ski. He'd decided to try it, and because it was the last opening Big Bear Lake had in the day, I knew I wouldn't be able to try flyboarding. This pissed me off. No, I didn't pay for it, but it was yet again another activity that I didn't get to do because I had supported the bike. It was frustrating. I didn't mind supporting BenHop, but I felt like I was missing out. So I sat on the boat the documentary crew was using to film and watched Jonathan's attempt while the rest of the crew videotaped it all. He had a good time.

After the flyboarding was over, we drove over to Big Bear City, where we had a consumer event planned. The event was being held at the Tourism Center, and when we pulled in, some of the center's employees were just setting up. We pulled up some chairs and shot the breeze that late afternoon with various living-green and alternative-fuel enthusiasts. We had a few hybrid owners show up, and people were surprised when I told them I had come up "the back way" to Big Bear Lake from Barstow; they knew how tough that climb was, and a couple of LEAF owners didn't think they wanted to gamble on making it up that way!

Finally, after the consumer event wound down, we drove over to the campground that Susan had set up for us to stay overnight in. We pitched our tents and rolled out the sleeping bags, and since it was a nice evening, we decided to head to downtown to eat and have some drinks. We ended up at the Big Bear Lake Brewing Company for dinner. Because it was already late, there was not a huge crowd, and Susan talked management into letting Dominique perform some songs. Everyone enjoyed the music, and, before long, Ben Hopkins sat down at the microphone and performed a few songs, with the highlight being his rendition of "Fish and Chips Lady"! (It must have been a British romance song?) Another late-evening good time.

Eventually, we headed back to the campground. I knew I would be going down the mountain in the morning and would regenerate power, so the battery wouldn't be a big problem for our drive. Thus, I plugged into a 110v outlet and let *Green Lightning* charge overnight on it.

CHAPTER 39

Day 36: Big Bear Lake to Riverside/Corona

"Downhill Glide to Riverside."

"Can't Wait to Be Late."

When we awoke to the fresh-air smell of pine trees and bright sunshine at Big Bear Lake the next morning, most everybody was in a good mood. Our ride for the day was to Riverside, about seventy-five miles away, and it was all downhill. We had an event with the city, which was pushing a "green" agenda, and the city had planned a town hall meeting, along with some invited hybrid-vehicle owners. There would be some press there, as well.

After we had packed up the camping gear, we rode over to a bagel spot in town, where everybody ordered breakfast. Benswing, Sean, and Jonathan were soon talking about doing something fun again, since we had some extra time. They were considering mountain biking, and something else. They talked Susan into it, and soon the group was headed that way. I had been dealing with a torn meniscus in my knee for the entire trip, so I decided not to join them. It was agreed everyone would meet at 12:00 p.m. sharp at the bagel spot to ride as a group to the town hall event.

With nothing better to do, I decided to go around Big Bear Lake and Big Bear City to take some pictures and to explore a bit. The only time I had spent any time in the area had been to snow ski. I was satisfied to spend my time relaxing by myself for a change.

Outdoors at Big Bear gave me time to think a bit, as well. I was perhaps

Exploring Big Bear with its beautiful lake and hiking opportunities.

feeling some anxiety about just having retired in haste to do the Tour. When I had done so, I had had a loose agreement with Susan to be the controlling executive manager of her electric scooter business. But it had become evident to me in our discussions that we were not going to agree on a couple of key issues.

One was about the best way to build brand image. Susan is an idealist focused on eliminating the middleman in distribution and selling directly to consumers via the Internet. I thought that her method would take far too long to build a brand and that the advantage which her scooter's power/design had currently would be lost far before a brand could be built. My idea was to use the existing scooter/motorcycle distribution network and sell as much as we could as fast as we could to build brand and volume.

The other issue was relative to customer service. Susan wanted to have one mobile mechanic or use other local mechanics to provide warranty and service. I felt this was risky and left customer satisfaction out of our control; and with a Chinese product, we could easily get a bad brand reputation if there were any quality issues that couldn't be fixed quickly.

To put it simply, my grand idea of growing a large electric scooter business was fading fast because our differences would not be ones we could overcome. There were other differences between us that I'd come to realize, as well. It didn't take much to convince myself that I had to walk away from our business agreement. I was disappointed, but I knew it was the right thing to do. I couldn't solve my next phase in life today, but I knew it wouldn't be running Susan's scooter business.

Green Lightning *enjoys the views and sunshine at Big Bear Lake.*

Hiking and biking opportunities abound at Big Bear Lake.

It was now approaching noon; I needed to get back to town. When I arrived back at the bagel store, I expected to see everyone. But no one was there. I sat around, waiting. Waiting and waiting some more. I tried calling others, but neither Susan nor anyone else answered. 1:00 p.m. came, and went. 1:30 p.m. I still had heard nothing from anyone. Finally, around 2:00 p.m., people started showing up. By now, I was pissed. I had wasted two hours just waiting on everyone. When I asked Susan about it, I got a really lame answer on why they hadn't returned on time. They couldn't get on the mountain right away and spent a lot of time filming those that mountain biked, I guess. But that didn't explain why they hadn't at least called to let me know. It was inconsiderate in my book. And we'd also been late—many times—to the folks who were setting up consumer events for us. My frustration was growing.

Susan decided that Benswing and I should go ahead of the group so that we could get to the Riverside town meeting event on time; the rest

Picturesque venues in the summer at Big Bear Lake.
www.bigbear.com/visitor-guide

Summer season at Big Bear Lake offers lots of activities
and picturesque venues for events like weddings.
www.bigbear.com/visitor-guide

would follow as soon as they could. Susan's plan sounded good, but it wasn't. Because, by now, it was nearly 2:30 p.m., and we would be going down a steep, winding mountain road where you could not make interstate speeds. And then we would be facing Los Angeles traffic in late afternoon, and who knew how bad that would be? We would be lucky if we made it there on time.

Benswing, Rachel, and I took off on a race against time. Had I known the others weren't going to return, I could have left at noon to get there for the meeting. But that didn't happen. Now we were winding around the tops of the mountain ridges along what's called the Rim of the World Highway at faster speeds than desired so that we could get to Riverside by 5:00 p.m.

We then picked up California SR 330, which was a better road, and even four-lane in parts, but the road was still windy and steep, and the traffic was heavy. It was more like a race course for downhill skiers! Nevertheless, we made it to the bottom of the foothills of San Bernardino quickly and soon picked up the highway for Riverside.

Even with all of our rush driving, we couldn't make the meeting at 5:00 p.m., as we'd planned; it was about 5:30 p.m. when we arrived. By then, one of the key city officials was growing impatient and had another commitment, so he greeted us before promptly departing. We did get to spend some time answering questions for the small group of city employees, local citizens, and press that had showed up. I think Benswing spent most of the time with the reporter. After the meeting, we went outside and met with a couple of hybrid-car owners. They were the "converted"—those

Rachel enjoys a short stop to view the Rim of the World views before we descend to LA.

The Rim of the World Highway is beautiful, but it is not a road you can make good time on.
www.myscenicdrives.com/drives/california/
rim-of-the-world-scenic-byway

that already were ready for alternative fuels. But the event could have been bigger perhaps had all the Tour members been there on time.

After about an hour, the event was over, and we headed to the hotel. I checked in and took the LEAF to a nearby Walmart Supercenter, which had a charger. It was a quiet night, and I had dinner by myself. I was also pissed because we had blown an opportunity for press and had let down our meeting coordinators all because some wanted to mountain bike for fun. It seemed that lately our mind-set had become we "can't wait to be late." It was frustrating.

The riders hadn't made it to the event and instead went straight to the hotel. I later learned that Jonathan had been riding a scooter for the first time and had taken a spill as the riders were descending the mountain grade. While he wasn't seriously injured, his arm was bandaged. I heard the ride down the mountains was hairy for all involved because the four-lane highway portion speeds were fast and traffic had little patience for the slow-moving electric scooters and BenHop's bike. But with no shoulder to the highway, there was no place to go, so drivers were flying by at fast speeds and honking their horns at Sean in the protective supply truck and at the riders as they went by. No respect from the California drivers! And ironically, most likely, those same drivers would descend to the LA traffic jams, and they'd be late, too. Just more "can't wait to be late" in LA.

Talking with some of the Riverside town hall meeting attendees and coordinator (left).

CHAPTER 40

Day 37: Riverside/Corona to Newport Beach

"Staking a Cross-Country Claim." "Screwed on the Beach."

Monumental fun. That's what today would be. With only a thirty-seven-mile ride, we would touch the Pacific Ocean, and thus be able to say we had completed the first part of our objective—i.e., to cross the country with all of our electric vehicles. And it was Friday, a good night to party on the beach!

I left around 9:00 a.m., with BenHop, and headed toward Orange County, a place I knew well since I had lived there for seventeen years, when Nissan still had its Sales and Marketing headquarters in LA. But to get to Orange County, we had to do some quick-skirting on the California SR 91 freeway—a highway known for accidents, high speeds, and stop-and-go traffic. This highway goes through a narrow canyon, and there is no alternate road to use anywhere. So we went as far as we could on side roads and then finally got onto the freeway. I had to let Ben go it alone and hope he didn't get caught, because traffic was simply too fast for me to drive on the berm safely. Since it was a relatively short distance (seven and a-half miles), Ben was able to cover it in less than half an hour without incident. But this was only part one.

Part two was to exit onto California SR 241. This was a more recent cut—through Santiago and Gypsum canyons—which saved traffic nearly an hour from going further west around the small mountains of Orange

County. If we didn't take this toll road cut through the canyons, driving to Orange County would probably take us an extra two hours, at least. So we had to risk taking the cut through.

Ben took the exit and started up the toll road. It was a long, uphill grade for the first couple of miles, and, midway up, I saw he was running out of power. Fortunately, I had lagged behind, and I pulled up to him with a fresh battery. The toll road wasn't nearly as busy as California SR 91, so stopping wasn't an issue here.

We continued on, watching for the highway patrol and the local sheriff. Fortunately, we skated by, and soon we were on the "backstreets." The word "backstreets" usually implies low traffic in most areas of the country, but, in Orange County or LA, it just means "not on the interstate." Most backstreets are four lanes wide, with plenty of traffic. Nevertheless, even though the streets were busy, we easily made our way further south and west toward the ocean, and our hotel in Laguna Hills. We had lunch along the way, took our time, and eventually reached the hotel around 4:00 p.m., the time when Susan had set up a meeting for us with some Orange County LEAF owners who were intrigued with our trip.

These LEAF owners were very proud owners, and they knew their cars. In fact, they had questions that I couldn't answer. Fortunately, Benswing had also arrived on the motorcycle, and he deflected some of the technical questions about extra kilowatt hour charging and such. The owners were quite interested in my power cords, which had allowed me to make it all the way from the East Coast. I could have sold some on the spot if I had had extras to sell! They were also fascinated that I had made it up to Big Bear Lake and other mountains, since they had never tried that themselves. It was a fun discussion.

Shortly after we arrived, a friend of Benswing's showed up. His name was Terry Hershner, otherwise known as "Electric Terry." Terry also rides a Zero electric motorcycle (similar to Benswing's), but it's modified with a Vetter front air dam and additional charging capacity to rapidly charge his motorcycle for cross-country riding events. An electric-vehicle advocate and public speaker, one of Terry's recent accomplishments is being the first ever electric motorcycle rider to go 1,000 miles in twenty-four hours, which earned him an Iron Butt Association award. Now that's an award to be proud of!

Terry was a bit of a showman. He loved talking to people about his motorcycle, electric vehicles, and what he does, which is mainly to ride that motorcycle all around the country to various electric-vehicle rallies. He talked shop with the LEAF owners, and they loved it. Between Benswing and Terry, the LEAF owners got all the tech information they could want. And that was at it should be, since Terry and Benswing are both "walk the talk" in terms of commitment to electric vehicles; they live electric.

After about an hour, the scooteristas, the documentary crew, and Sean arrived with the supply truck. We quickly unloaded our luggage at the hotel and then headed out to Newport Beach, where Jonathan wanted to film our arrival as we touched the Pacific Ocean. With a few of the LEAF owners tagging along, we headed down to the Pacific Coast Highway and made our way north to Newport.

When we arrived at the Balboa pier area, we had to talk the local police and business owners into letting us park our parade of vehicles along the popular boardwalk. Our idea was that we would talk to passersby about our trip and electric vehicles while we were there.

So we parked, and then, with cameras rolling, we all ran into the ocean to celebrate! And why not? After all, we had just accomplished what no one else had: we had driven an electric vehicle, actually four *different* types of electric vehicles, across the vast United States! It *was* something to be

"Electric Terry" talks about his Zero electric motorcycle
to LEAF owners at our hotel.
https://en.wikipedia.org/wiki/Terry_Hershner

proud of.

After that, there were a lot of silly kids games on the beach. Jonathan got plenty of video, and we gradually made our way back to the boardwalk and mingled for a while with passersby. We were a collective happy and proud Tour. We'd made history today, even if nobody knew it.

Later, we started our party by eating dinner together at the Mexican restaurant, Cabo Cantina, which was right on the boardwalk. Before long, the drinks were flowing, the toasts were growing, and we partied for two or three hours. Terry was there. Some of the LEAF owners were there. People we had just met were there. It was a great time had by all. The only thing missing, of course, was the press.

Eventually, we had to wrap it up and head back to the hotel. I went out to the LEAF, which was parked along the boardwalk, and was talking to a security guard when I noticed my left front tire was a little low. "Strange," I thought. I checked it over, and, lo and behold, there was something in the tire which appeared to be a screw of some type. Great. Just what I needed at eleven o'clock on a Friday night.

But the tire wasn't flat, so I decided to chance it and drive back to the hotel as is. I didn't know how long the screw had been there, so I'd just hope it wouldn't blow for the ride to the hotel tonight, and I'd get it fixed in the morning. So with that, we all packed up and headed to the hotel, and, luckily, my tire held up.

I didn't charge the LEAF this time. I'd worry about that, and the tire, in the morning.

A celebratory photo of the Ride the Future Tour as they make it to the Pacific.
www.cabocantina.com

CHAPTER 41

Day 38: Newport Beach to Santa Monica

"A Customer is a Customer X 2."
"Route 66 Does Re-a-Pier."

We had made it across the country in thirty-seven days, which was pretty good time, considering all of the cities we had visited along the way. Our journey was not a race to see how fast we could get to the Pacific, although it often felt like a race since we never stayed in any location for more than about twelve hours. It was a trip of endurance and distance. And it wasn't quite over. We now turned our attention northward, basically following the California coast all the way to San Francisco. Using the Pacific Coast Highway (California SR 1), we would see some beautiful scenery and finish our ride by crossing the Golden Gate Bridge, before ending the Tour at Google Headquarters in Mountain View—that is, if Google let us in.

The ride on this Saturday morning would be an easy one, only fifty miles from Newport Beach to Santa Monica. This stretch of the Pacific Coast Highway (PCH) has some of the prettiest and busiest beach scenery in the country, so traveling it would be a great opportunity to showcase our electric vehicles and tout our accomplishment along the way. But, sadly, I would not be able to join the Tour on the ride; I had two more urgent matters to deal with: a low battery and a low tire.

I left the others around 9:00 a.m. and headed for Irvine Nissan, the closest Nissan dealer, so I could recharge the LEAF. When I got there, I found that it had a newly installed CHAdeMO quick charger. Quick chargers are great for electric-vehicle owners because they can charge using 440v power, which reduces charging time from four hours for a full charge at 220v to less than thirty minutes. While quick chargers are becoming common at dealerships today, in 2013 there were not many around, so this was sweet for me. It would be a lot faster to recharge than I thought.

www.evsolutions.com/ev-charging-products-for-business

I waited for a customer to finish charging ahead of me and when the person left, I pulled the LEAF up to the quick charger and plugged in. I had just started when a couple pulled up in a LEAF, and we briefly spoke before I decided to go inside the office of the dealership. Once inside, I introduced myself to one of the salespeople and let the person know that I was setting a Guinness World Record, but I didn't mention that I was a former Nissan corporate employee. I sat down and was doing e-mails and texts. A few minutes later, I noticed the general manager walking out to the couple outside, and I heard him asking them if they were waiting to use the charger, and they replied they were. I decided to walk out to the LEAF to see if it was done charging, which it basically was, though it couldn't have been charged for long. But as I walked out, the general manager advised me that I needed to move my LEAF and that the charger was for retail customers.

This immediately made me mad because 1) I didn't know the charging had finished and if the couple had asked me to do so, I would have quickly moved out of the way, and 2) if I wasn't a customer, what was I? I quickly apologized to the couple and unplugged from the charger and moved the LEAF, while the general manager went back inside the office. At this point, I had a decision to make: do I confront the general manager, or not? From my perspective, he didn't know who I was, but had I lived in the area (which I had seven years prior), he would have lost me as a customer or potential customer forever. I *was* a *Nissan* customer, whether I bought/ leased/borrowed the car from Irvine Nissan or not.

This narrow-minded thinking has been an issue for dealerships forever and was a pet peeve of mine. On the other hand, dealers can do what they want because they are businessmen, and confronting him as a retiree would probably have just escalated into an argument neither one of us could win, so I just left. Sometimes I avoid confrontation, and I walked away from this one, right or wrong. I didn't even take a picture.

A short distance away was a tire dealership. I drove over to it, and walked in. The technicians put me on the list because they were busy on this Saturday morning, but they assured me they would get to me as soon as they could. It was well over an hour and a-half before they could get to my tire, so I sat around trying to make the best use of my time.

As I sat and waited, I was a little worried about what this might cost me; if I needed a new tire, it wouldn't be cheap. I had purchased a spare wheel and a Michelin tire as a spare for the trip at a cost of some $400, but I was hoping the technicians could simply repair the one I had for a few bucks. It would save me from using the new wheel, which I would sell when I got home.

Finally, the head technician came out to greet me and stated that the technicians had removed the screw and that, fortunately, the hole was not too close to the rim and had not done any other damage, so the technicians were able to easily plug the hole. No new tire. Just a plug! This was great news! I asked him what I owed, and, to my surprise, he stated that America's Tire's policy was that it didn't charge for tire repairs, so I didn't owe the business anything. America's Tire just wanted me to come back to it whenever I needed new tires. Really? Nothing? What customer service! I was so happy I asked my technician to pose with me for a picture and was sure to get America's Tire brand name in the picture! And if I ever do need tires (which for me would be rare since I lease from a car company), America's Tire will be my first choice. I was certainly impressed with its approach to business.

My technician at American's Tire made me a happy customer by repairing Green Lightning's paw.
www.discounttire.com

So to summarize my early Saturday morning:

Two hours.

Two customer interactions.

Two very different results.

Now I was in a good mood, and I headed toward Santa Monica to meet up with the Tour again. However, since it was lunchtime, I decided to stop off at Duke's for a good hamburger and beach "scenery."

Duke's is an iconic restaurant located right on Huntington Beach. It's named after Duke Paoa Kahanamoku, the legendary Hawaiian swimmer and surfer who spent many years in the water along the Los Angeles and Orange County beaches. His life is an amazing story, and includes

- Winning five Gold and Silver medals in swimming over a twenty-year Olympic period.
- Playing parts in twenty-eight Hollywood movies while living in the LA area.
- Rescuing eight drowning fishermen with his surfboard on the same day in Newport Beach.
- Being reelected thirteen times as sheriff for the city and county of Honolulu.
- Being the first person inducted into the Swimming and Surfing Hall of Fame. http://www.dukeshuntington.com/duke

Duke was instrumental in the expansion of surfing as a sport worldwide.

Now when I go to Duke's (the restaurant), I do enjoy the excellent food and outside bar, but I also enjoy the "scenery." You see, Duke's is in the heart of the beach life. Thousands visit this beach each week. So there is always something going on; it could be anything from kite flying to surfing to volleyball tournaments. Sometimes on Sundays, I would drive up to walk the pier and then watch the crazy group of bongo drummers and dancers who gather at 3:00 p.m. to perform for passersby; there is no leader—it just happens.

I had my burger and enjoyed the "scenery" of the day (today it was just a few hundred pretty women in bikinis) and then took off for Santa Monica again. I rolled through more beach communities—i.e. Sunset beach, Seal Beach, Alamitos Beach, Long Beach, Redondo Beach, Hermosa Beach, Manhattan Beach, and Venice Beach, before finally reaching Santa Monica Beach.

The rest of the Tour had made stops along the way, including a visit with Paul Scott, a well-known electric-vehicle and solar-power enthusiast and cofounder of Plug-In America (https://pluginamerica.org). But we all were due to arrive around 4:00 p.m. because Jonathan wanted to film on Santa Monica Pier. Benswing was off to swing dance with the local LA group of dancers, but the rest of us would meet up on the pier.

When I arrived, I drove directly to the pier because my planning had shown that the city of Santa Monica had placed rechargers right on the pier itself, which was unique. Not only unique, but convenient. If you're familiar with Santa Monica, you know that parking is a premium during the summer months due to the huge crowds. But if you drive an electric vehicle, you can park on the pier and have access to all the activities, with no walk to get to them!

I got to the pier, and, sure enough, the pier parking was packed. However, the electric-vehicle charge spots were not! So I made my way to them, parked, and plugged in to recharge. Of course, the first charger I picked out didn't work—again! However, the second one did, and I was golden.

I looked over these chargers because they were new to me. Across the United States were two main public-charging networks: Blink in the East and the Midwest, and ChargePoint in the West. (www.chargepoint.com) These public chargers were apparently owned by the city and not part of either network; they were free to use. And the chargers were not from AeroVironment, which makes most of the public chargers at Nissan deal- erships. These were made by Clipper Creek, another charger supplier. But what caught my eye was the first one in the line of chargers.

A postcard placement for an electric vehicle charger—
on the beach in Santa Monica!
www.clippercreek.com

The first one in line was not the typical Level 2 charger with a J1772 connector for charging; it was a charger made by MagneCharge with a J1773 connector. This charger was unique in that its charge port used an inductive charge paddle, of which there were two sizes, a small paddle and a large paddle, which are inserted into the vehicle. The system was designed to be safe even when used in the rain and was used on early GM electric vehicles, like the EV1 and Chevy S10 EV, as well early Toyota RAV4 SUV models. It is nearly obsolete today, having been replaced by the J1772 connector industry-wide. It did make sense to see the obsolete charger here in Santa Monica, because stringent California emission laws had made the state a test market for all early electric vehicles. To the city's credit, it was an early adopter in supporting electric vehicles and having public chargers. Kudos are in order to the city for being out in front on vehicles without emissions!

I looked around for a bit. This location was ideal, right on the pier itself, which had a small amusement park in the middle of it. Not only could I watch the beach players but I could mingle on the pier. I made a call and learned that the others would be along shortly and that we would meet on the pier.

I took a couple of pictures of the beach activities and waited. It wasn't long before we were all together and walking the pier as a group. For Rachel and BenHop, this was their first visit to a Santa Monica, so the sights were fascinating. Even if you have been to the pier before, it's always fun to take in the sights and sounds; it's one of the finest beach environments in the United States. And with no evening event planned, we took

The city of Santa Monica has several Level 2 chargers on the pier, including an old paddle-style charger.

The Santa Monica Pier has a small amusement park and
electric-vehicle chargers.

our time watching the singers, the dancers the parasailers, the beach bathers, the amusement riders, the gamers, the surfers, and more.

Jonathan's team took more video at the end of the pier, and then we sauntered back toward the beach. We paused to reflect at a symbolic marker. That marker was a former tollbooth/gate-keeper booth that had been converted to a retail shop. However, it was also a billboard marking the end of Route 66. We had lost track of Route 66 when we detoured northward to Hoover Dam and Las Vegas. However, we had actually crossed it unknowingly as we made our way south into Orange County.

But here it was again, fittingly bidding us farewell at its terminus. We had spent many days following its historic past across the United States, picking it up in Oklahoma and now visiting the end of it in Santa Monica. It was bittersweet. Like leaving an old friend. We took a few pics with the signs to memorialize the moment and left Route 66 to romance the next traveler that came down her path.

The sun was descending, and it was nearly time for dinner, so we headed over to the many restaurants that Santa Monica has to offer. BenHop had

Santa Monica beach activities abound.
http://santamonicapier.org

The sights and sounds on Santa Monica pier with the beach and sea below.

some friends in the area, and he invited them to join us all for dinner.

Someone picked a nice restaurant, and we all tried to get in, but because it was a Saturday night, we got split up. I had dinner with Ben and his friends, who were very nice people; we had a great California tofu-type dinner and pleasant conversation before I decided it was time to head for the hotel.

Rachel and I had kind of chummed on the pier, and she joined me as we headed back to the pier to get the LEAF. Rachel was enamored with

Santa Monica—it was like she had found her home! I didn't know if she was going to leave in the morning she liked it so much. But isn't there something about a "rolling stone keeps on rolling"? Maybe not, but it would be appropriate to describe Rachel.

Amidst the crowd, the Route 66 sign stands out.

The Tour pauses for a group photo at the end of Santa Monica pier.

Rachel gushes about Santa Monica, but the LA Party is over.

CHAPTER 42

Day 39: Santa Monica to Santa Barbara

"Where and When Will We Find Ben?"

"No Candelabra in Santa Barbara."

I live in a perfect world.

It's not dark, like *The World According to Garp.*

It's not a fantasy, like *Wayne's World.*

But it is *Duane's World.*

And when my perfect little world doesn't line up with the outside-world reality, I tend to absorb my frustrations and eventually lash out in some way.

It's happened in my work life.

It's happened in my personal life.

And today, it would happen on the Tour.

Our objective for the day was to make it to Santa Barbara, a lovely coastal community about ninety miles north of Santa Monica. We would be leaving the LA metro area today and follow the Pacific Coast Highway (PCH) along the coast all the way to San Francisco from here. It's a beautiful drive, especially if you do it at the speed we would be doing it, a speed allowing us to take it all in.

Rachel was riding with me again, and, after our morning briefing with BenHop, we headed back over to Santa Monica pier to "top-off" the LEAF and look around more. It was an overcast, coastal-fog, gray morning, and few people were around early on a Sunday. We walked around the north side of the pier to check out the large indoor merry-go-round and kid's playground. Sadly, the merry-go-round wasn't operating yet, or I'm sure Rachel would have ridden it.

We also found a display on the beach that we hadn't seen yesterday—a display of red and white crosses. Curious, we walked down to it. It turned out that this was the *Arlington West* display, a project designed to draw public attention to the subject of military and civilian deaths during the invasion and occupation of Iraq.

Designed by local activist Stephen Sherrill to be a place to mourn, reflect, contemplate, grieve, and meditate; to honor and acknowledge those who have lost their lives; and to reflect upon the costs of war, it has been adopted by the local chapter of the Veterans for Peace. It is reinstalled

The indoor merry-go-round at Santa Monica
pier is a novelty for kids of all ages.
Below: The kids' playground includes unique "sand" sculptures.

The Arlington West *display on Santa Monica Beach.*
https://en.wikipedia.org/wiki/Arlington_West

each Sunday and Fourth of July by volunteers and has been since its inception in Santa Barbara in November 2003. It is a haunting reminder of the high costs of war.

After paying our respects, we made our way back to the LEAF and started up the PCH. We passed the many coastal residences and communities hidden in the canyons and cliffs of the east-side hills, as well as the beaches that extend continually on the west side of the PCH. And then we moved into the area where the mountains reach the ocean, passing famous landmarks like the Getty Villa, Gladstone Restaurant, Topanga and Carbon Canyon roads, and then into Malibu. Shortly after leaving Malibu, we spotted BenHop and drove past him. We stopped at a nearby pullout to check out the coast and were expecting Ben to see us and stop, too, but he was so focused on riding that he went right by us!

We continued on, passing Ben again and stopping at a roadside fruit stand near the city of Oxnard. The Oxnard area has a very large

Here comes Ben. And there goes Ben.

migrant-worker community and is one of the largest strawberry producing areas in the country. We bought a few strawberries and other fruit and were on our way. We hadn't seen Ben go by us, so we assumed he was still coming along the PCH. However, after waiting awhile and not seeing him, we weren't so sure he was. Then we decided to try to see if he was ahead of us, and we drove further north, but we soon realized we would encounter a problem because our destination was the point where the PCH (California SR 1) meets U.S. 101 (the Ventura Highway), a major north-south four-lane highway with narrow lanes, limited roadside berm for slower vehicles to use, and lots of fast traffic.

The Santa Monica Mountains meet the ocean near Malibu.

At this point, we didn't know which way to look for him. Of course, we could just call him and have him meet us somewhere. Wait! This was Ben, and he still hadn't replaced his phone, so we couldn't call him. We would have to wait for *him* to contact us. Ugh. So we pulled onto the next exit of U.S. 101 and waited. It was all we could do.

BenHop did eventually call. He had missed the point where the PCH split from Oxnard Boulevard, and he'd ended up in downtown Oxnard. That meant meeting him at a gas station and giving him a fresh battery. We sent him on his way, and since we needed to recharge, Rachel and I drove to Team Nissan in Oxnard. There we plugged in and walked to a restaurant in a nearby shopping plaza and had lunch.

After we picked up the LEAF, we went searching for Ben again. But U.S. 101 continued to pose a challenge. There were stretches along the coast where the mountains were so close to the ocean that U.S. 101 was the only road to travel on. Those stretches were not fun for the riders. In fact, probably nerve-racking is a better description. Fast traffic and narrow roads are not any rider's idea of a good time.

There were also stretches where there was a frontage road or a small, two-lane coastal road through small towns, which provided access to beaches, fishing holes, and surfing bays. And there were stretches where the PCH (California SR 1) was separate from U.S. 101. But just as easy as the routes separated, they would quickly merge back together. That made it difficult to stay with Ben and almost impossible to know where he would go. He didn't have a detailed map. He didn't have a cell phone. And as we'd just witnessed, he *did* have tunnel vision at times!

Charging the LEAF at Santa Barbara harbor.

We kept driving. Then stopped and waited. Just when we had just about given up on finding him, suddenly there he would be, still cycling away, mile after mile. It was frustrating, but at least we never totally lost him—though we probably should have.

We finally made it to Santa Barbara, and we were actually early. The Tour had a planned meeting with the Santa Barbara press at 5:00 p.m., so I decided to go a little further along the marina area to find a public charger that was supposed to be there. Rachel and I hung out for a while as the LEAF charged, before heading back to the parking lot where we were to meet the press.

When we got there, the other riders from the Tour hadn't quite arrived yet. So we chatted with a newspaper reporter and a cameraman until Susan and the others came in, with horns blaring and hands waving.

Soon the cameraman wanted to get film of the scooters and bike. Later he had me reenact driving into town so he could film it, which I did. After a couple of drive-bys, he had what he wanted, and I returned to the parking lot, where members of the press were about to do the interviews. They had already gotten some quality time with BenHop. Now they wanted to get more background on the Tour and presumably the Guinness World Records we were trying to set.

Jonathan and his team arrived, as well, and quickly engaged with the reporter. Jonathan was now taking charge, and he convinced Susan to let the girls—Dominique and Rachel—be the ones interviewed about the Tour, since this would be a TV video interview. And so that's what they did: "the girls" spoke about the Tour's purpose, what vehicles were involved, and their roles in the trip.

After that, the interviews were over. The press had my name and where I was from and such, but no real formal interview. With Jonathan now making the decision, I was not a part of one of the few formal press events we had had. This didn't sit well with me. In fact, it didn't sit well with me at all.

The group then went to dinner at a restaurant close by. The meal was okay, but I wasn't very talkative. I couldn't let go of my disappointment and frustrations. I stewed over them during dinner and couldn't let them go. My frustrations had been building for some time, and, now, I was mad.

The press event raised the question again about what our priorities were. Way back in Tennessee, Susan had asked the group if we all knew what our objective was, and none of us had said anything. Yet I don't think there was clarity, because, in part, Susan had many agendas—from setting Guinness World Records to educating the public about electric-vehicle alternatives to gasoline, to promoting sustainability, to building her (and my) brand image for the Xenon electric scooters, to having an award-wining documentary film and even to giving Dominique exposure to launch her solo music career. They were all in there.

To me, this had been our only major opportunity to be featured on TV to promote what the Ride the Future Tour was all about. I thought it

The beauty of the Santa Barbara marina and beach area.
www.santabarbaraca.gov/gov/depts/waterfront

was about setting Guinness World Records and, as such, should feature BenHop, Susan, and me. Possibly Benswing, too, although he was pretty sure that someone else had already eclipsed the distance record that he was trying to set.

Instead, it was Dominique and Rachel who did the interviews, neither of whom was setting any records—unless you're including record albums. Moreover, they were the two people who did the least in terms of preparation and work to make the Tour successful. It wasn't a matter of not appreciating their contribution and support efforts, but more of where the focus should be for one of our few PR opportunities. But that wasn't all of my frustration. That Jonathan had made the decision also bothered me. Jonathan and Susan had gotten a lot more friendly lately, and he was having more influence on our daily activities. The relationship didn't bother me; his influence did, because some of the decisions were more about making a better movie than getting the basics done. Maybe I felt I should be getting more documentary face time. Or maybe I was still pissed about the Big Bear Lake water adventure and other fun activities I had missed out on.

I'd been sharing some of my frustrations with Sean, my usual hotel roommate. I'd been asked by Susan to support BenHop on the bike, and I had done that willingly even though it meant a lot more work for me to provide an ongoing supply of batteries to him every day. I was still navigating the route every night (though nobody appreciated it), managing charging locations for the LEAF, as well as managing Ben's batteries. I was even fine with all that. But when there were PR opportunities, I wanted to be involved. After all, I was trying to set a Guinness World Record. Instead, what had been happening was either 1) we were late getting to the event and missed the PR opportunities, or 2) Benswing and/or Susan got there earlier than I, since they could go faster, and did the interviews. Benswing, in particular, was a media sponge and ensured he got media coverage when and where it was available. To his credit, he had the technical background to give the press what they wanted to hear, but we were supposed to be a team.

Being available for the press in Santa Barbara, finally, and not getting the opportunity for an interview became a final straw for me. I walked around the pier after dinner, asking myself why I was doing this if not for 1) the future job that now wasn't going to happen and 2) for the press and exposure for setting a Guinness World Record. I needed to think.

I told the group I needed to finish recharging the LEAF, so Rachel caught a ride with someone else. Susan had done a house swap with a family in Santa Barbara, so the Tour was staying overnight in a house up on one of the hills in the city. Everyone headed over to the house a few miles away as it grew dark. Everyone, that is, but me.

I did go down to the marina and charge the LEAF. But I also sat in the car and debated whether I wanted to finish the Tour. I had left my job feeling frustrated and underappreciated, and, now on the Tour, I was feeling the same way. I knew though that leaving the group wouldn't get me much, either. I stopped taking phone calls and decided to sleep on it. I would decide in the morning what I would do. I knew I'd be letting Susan know I wasn't happy by not showing up at the house, but I didn't care. I was, in my own way, "lashing out" at the real world from Duane's World.

The LEAF is a wonderful car with a lot of features and benefits. Sleeping in it, however, is not one of them. I spent the night trying to sleep in an abnormal position in a small car in a dark parking lot. I don't know if I was more worried about getting robbed or about being jabbed in my lumbar region, but, suffice it to say, it was a very long night.

Stewing, while I took pictures around the pier at dusk.

CHAPTER 43

Day 40: Santa Barbara to San Luis Obispo

"Mountain Climb to Danish Time."

"I Can't Stop Wine-ing."

Waking up in a car in the morning, is like tent camping on the side of a mountain:

1. You're not fully rested because your body could never get comfortable;

2. your body hurts from odd positions and penetrating objects underneath you;

3. you feel dirty, and you stink, from sleeping in the same clothes you wore yesterday.

Nevertheless, I had calmed down from yesterday's stew-fest, and I decided to finish out the Tour. After all was said and done, I had invested so much time and energy into it that it simply didn't make sense to walk away. Yes, I was still frustrated for multiple reasons. But I would stick it out—and make the best of it.

I used the public restroom at the marina to clean up and change clothes. I was hungry and knew Santa Barbara had few fast food locations, so I walked across the street to a restaurant I didn't know still existed—Sambos.

Sambo's was a restaurant chain started by two friends, Sam Battistone and Newell Bohnett in 1957. They pulled Sambo's from a combination of letters from their names but soon found the restaurant associated with

The original, and the only remaining, Sambo's restaurant across from the marina in Santa Barbara.
https://sambosrestaurant.com

the 1899 children's book, *The Story of Little Black Sambo*. Recognizing an opportunity, they decorated the walls with scenes from the book, including a dark-skinned boy, tigers, and a magical unicycle-riding man called "The Treefriend"; these were later modified to fit the restaurant brand via a kids club.

At its height, there were 1,117 Sambo's restaurants in forty-seven states in the United States, before racial tension and lawsuits led to its quick demise in the early 1980s. Now there was only one left, the original one in Santa Barbara, and I was in it.

Sambo's was, and is still, a throwback to the 1960s diner atmosphere. Only open from 6:30 a.m. to 3:00 p.m., it specializes in breakfast and lunch, and it does it well. My breakfast was tasty and filling, and I wondered why the restaurant ever went out of business. Too bad.

After leaving Sambo's, I looked at my watch. It was still early, only 7:30 a.m., but I knew our leg was another long one (some 110 miles or more) to San Luis Obispo, so getting started early was important. I drove over to the house where the Tour stayed overnight and went inside. Almost everyone was still asleep, and, in fact, I had to wake BenHop.

While he got dressed and ready to ride, I loaded some recharged batteries into the LEAF for his bike. I decided to check out the newspaper I'd picked up off the driveway on the way in and read it while I waited on Ben.

And when I did, lo and behold, there I was, front and center, on the front page of the *Santa Barbara News-Press*! Along with a nice article about the Tour and its objective to set Guinness World Records, this was obviously the best press we had gotten so far.

I later heard that the girls' interviews did air on one of the TV stations also. Maybe our momentum would pick up now that we were getting close to the end of our journey. Maybe we would get more press. Hell, maybe Google would even let us in the gates at the end! I felt a little guilty for my thoughts and actions last night. Was I being narcissistic?

BenHop was ready, and we headed out the door, while leaving almost everyone else still sleeping in the house. I outlined our plan to Ben for the day and noted that we had two route options:

> ➤ one that led us along the coast and the busy U.S. 101 or

> ➤ a second that contained fewer miles but went through the San Marcos foothills—meaning that Ben would be climbing some mountains.

BenHop opted for the latter, since it would ultimately mean fewer miles for him to travel. If he was game, I was game for that route.

The route followed California SR 154 through the hills and was otherwise known as the San Marcos Pass Road. As we turned onto this road, we encountered some coastal fog, as well as signs telling us to turn on our headlights for the narrow and very curvy road ahead. Before long, we were slowly climbing into the foothills behind Santa Barbara.

Some of the best press we got was in Santa Barbara.

While not nearly as steep as some the mountains he had already climbed, these were still significant climbs. The electric battery power aided Ben immensely, but it still took his legs to make it up to the ridge-tops. Ben's ability to keep pedaling up these mountains was a testament to his true athletic ability and sheer determination. During the entire trip, I don't recall one time that I saw Ben walking his bike uphill. He took on mountains like a dog on a rawhide bone, gnawing away until there's nothing left. It was impressive to watch and even more impressive to see such athletic ability and sheer determination come from someone with such a positive outlook.

As our elevation grew, so did our views. Now, instead of the ocean coast, we had inland views of the Santa Ynez Mountains and the Santa Ynez Valley. And they were great views indeed! While there were still dry mountains, the vegetation was still more plentiful and green than those in the LA or the eastern Mohave area we had gone through. And as we rode along the ridgetops, we could even see some lakes.

Up and down we went, along the ridgetops for a while before we finally stopped at the vista point overlooking the Santa Ynez Valley. Ben needed a break, and it was a good place for one. The vista overlooks Lake Cachuma, a popular fishing, boating, hiking, and camping county park. However, because of its original development with Bradbury Dam, the lake is also a major water source for the area.

Starting our climb into the foothills of San Marcos Pass.

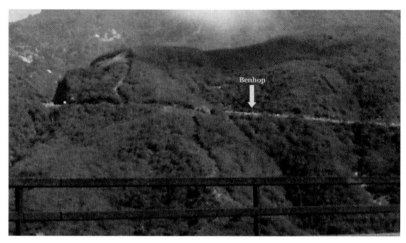

*Seeing the road ahead, I waited for Ben to
climb the next ridge to take this picture.*

Soon after the vista point, we began descending and quickly made
our way along the Santa Ynez River and past Lake Cachuma. We passed
numerous camps, marinas, and recreation areas, which looked like fun,
but not for us today. The land leveled out into farmland, where we came
to the town of Santa Ynez. Here we turned west onto California SR 246,
which took us into the town of Solvang.

I had been to Solvang before and wanted Ben to see it. Solvang is a
Danish town and the self-proclaimed "Danish Capital of America." A
small town of 5,200, it was established in 1911 by a group of Danes who
wanted to escape the winters of the Midwest. Most of the business buildings in the town reflect the architecture and heritage of the Danish culture. There's even a statue called *The Little Mermaid*—just like the one in
Copenhagen. The town is quaint; the town is cute. And, of course, it's a

*Views of the
Santa Ynez
Mountains.*

303

The greener vegetation of the Santa Ynez Valley, with Lake Cachuma in background.

Ben crests another hill on the narrow, windy San Marcos Pass Road.

tourist trap. But it's still interesting to visit.

I located the public electric-vehicle charger and plugged in. Ben parked his bike, and he and I walked around town. We found a little pastry shop called Olsen's Danish Village Bakery and had ourselves some coffee and a sweet. It was a bright, sunny day, and we sat outside and relaxed. It felt good. And maybe the pretty waitresses delayed our departure, as well. But I digress. Again.

BenHop and I walked around a bit while the LEAF charged, checking out the brewery and a few of the shops downtown. And then we hit the road again.

As we continued northwest on California SR 246, we got into more fertile croplands and, in particular, wine country. Santa Barbara and San Luis Obispo are both well known for being big wine-producing areas. In fact, San Luis Obispo County is the third-largest wine-producing area of California, behind the Napa and Sonoma regions. We passed several

Since BenHop didn't have a phone (surprise!), I took this selfie with my phone at the vista point.

A view of Lake Cachuma from the vista point.
https://countyofsb.org/parks/cachuma.sbc

wineries along our route, and since I was bored (driving 20 mph), I eventually stopped at one of them, did a couple tastings, and bought a bottle of wine. I'm not a big wine drinker, but it just seemed like the thing to do!

Meanwhile, Ben kept chugging along, and we passed through the city of Lompoc, where we picked up California SR 1 again, the PCH. And along the PCH, we would pass Vandenberg Air Force Base—a space and missile testing base—and eventually reach our destination for lunch,

Around Santa Ynez, we found farms with neatly arranged bales of hay.

Green Lightning *reloads at the public electric-vehicle*
chargers in the center of Solvang.
http://cityofsolvang.com

which was Santa Maria.

We grabbed a quick bite in Santa Maria and headed toward Pismo Beach, passing more wineries and crop fields along the way. It was fascinating to see how these huge farms could raise and manage such large fields, using large equipment and state-of-the-art technology. Big farming is big business, and it's in full display in central California.

We met up with the rest of the Tour in Santa Maria. I stuck pretty close to BenHop, which wasn't too difficult since we were on the PCH. I made

http://solvangbrewing.com

As the restaurant in yellow says, Solvang offers a "Bit of Denmark."

occasional road chat with him about sights along the way. And despite never having a phone and being a bit disorganized, Ben was a loveable character, and I enjoyed his company. He had a sense of humor, too, and was up for the occasional joke or crazy picture, just for fun. It was this personality and humor that gave me a sense of responsibility to support him and the reason why I did so. We were a pretty good team even if we were both independent. A big, fun-loving guy who enjoys life, from singing to beer to his making fun in the cartoons that he draws, that's Ben Hopkins. That's why people like being around him, and that includes me.

We moved on along the Cabrillo Highway (PCH) and came into Pismo Beach, which was back along the Pacific Coast again. A beach city, Pismo

One of the numerous wineries along our route to Santa Maria and San Luis Obispo.
https://dierbergvineyard.com/our-story/overview

While old and new met at this winery, it didn't have the "juice" Green Lightning needed.

Beach is a big recreation location for California residents, known for Dinosaur Caves and giant beach sand dunes where ATVs and Hummers rule.

Unfortunately, we didn't have time to stop, but passing by the signs made me want to pull over and take a Hummer ride. It's a little scary, because you can't see other drivers or riders, but it's still a lot of fun!

At the north end of Pismo Beach, the PCH merges with U.S. 101. We took a feeder road called Price Street and then Shell Beach Road, both of which paralleled U.S. 101. However, the road narrowed as it turned north, and we both ended up on the U.S. 101 again, which is not what we wanted.

Since I couldn't stay with Ben, I went ahead of him and waited. But he never came by. That's because once again he got stopped by the California Highway Patrol and was thrown off the freeway (U.S. 101), and then he subsequently got lost. Fortunately, he was near San Luis Obispo (SLO) and was able to find his way to the hotel easy enough. Just another day for Ben.

I finally arrived at the hotel around 6:15 p.m. and checked in. I needed to recharge the LEAF, so I looked up the only electric charging station in SLO, which was located three miles from where we'd just come—on the south side of the city. I drove over to the shopping plaza, located

Just one of the many large farms and crop fields we passed in the San Luis Obispo County area.

Ben was anxiously waiting for the movie to start—until I told him there were no cartoons.

the chargers, and looked around for a place to eat, which ended up being a place called Tahoe Joe's. It was a nice steakhouse with a bar, and a good place to spend a couple of hours while the LEAF recharged (https://tahoejoes.com). Benswing showed up so he could charge his motorcycle, and later Sean came by, and we talked for a while. I didn't see Susan that night, so we never talked.

After the LEAF was fully charged, I went back to the hotel and went to bed; I needed to catch up on sleep that I didn't get the night before. But before I went to bed, I had to make a phone call......

A local rock 'n roll diner housed in old passenger train cars is a sight in Pismo Beach.
www.pismobeach.org

A view back toward the Pismo Beach pier.

CHAPTER 44

Day 41: San Luis Obispo to Big Sur

"The LEAF Family Shows Its Colors."

"Seal-ing the Deal Along the Coast."

"To Sur with Love . . . But . . ."

Today was my Achilles heel. I had flagged this leg as being the one place where I could lose my goal to never use the backup generator. It may have been a secondary goal (the first being the distance traveled for the Guinness World Record), but it was a goal, nonetheless, because if the LEAF ran out of power, all of the electric-vehicle naysayers would have a field day with the fact that the car couldn't go everywhere and still wasn't a practical alternative to gasoline engines. Unfortunately, though I had spent hours looking for and calling places in the small town of Cambria, where I would need to recharge, I couldn't find any place that had access to 220v power. No hotels. No RV parks. No public chargers. And, quite frankly, the towns further north were so small that there were no options that I knew of. I was stumped.

Nothing that I could see on a map or on the Internet—and, mind you, this misfortune came after I had learned in Brinkley, Arkansas, that air conditioners might be one source—appeared to be available for a power source. In the middle of our trip, I talked this over with Susan and explained my dilemma. She had many connections to the electric-vehicle groups because she planned the consumer events that she set up. She decided to put out the word to the electric-vehicle and LEAF communities online and see if anyone had any ideas. And, it just so happened, someone did.

This leg of our journey took us from San Luis Obispo, which was slightly inland from Morro Bay on the coast and which followed the Pacific Coast Highway immediately adjacent to the ocean all the way to Big Sur State Park, our destination. It's an absolutely gorgeous drive, probably one of the most scenic in the country, with its blue water beaches on the west side and the base of steep mountains on the east side. The total distance was a long 110 miles, with the last half of it almost all uphill to Big Sur. And, in this latter half of the drive, there were no major towns, in fact, barely any towns. To further complicate matters, the only town after Morro Bay (Cambria) was only thirty-three miles from San Luis Obispo, meaning that my recharge would have to take me some seventy-seven miles or 70 percent of the distance to Big Sur and I would be going uphill. The conditions were ripe for failure. A major risk, and I knew it. And the risk assumed I would find power in Cambria.

This was such a risk that I considered going straight up the U.S. 101 from San Luis Obispo to San Francisco. It was a direct shot, and there would be easy power along the way. But the Tour had to use no major roads, and Susan wanted the group to see the PCH at its best, which I understood. I wanted to stay with the Tour if I could, so, early on, I decided I would take the gamble if I could find a power source. Would I make it all the way to Big Sur? I didn't know.

My call last night had been to Steve Schlopp. Steve was a LEAF owner who had answered Susan's plea for an "assist" with the electric-vehicle community. It so happened that Steve had recently moved and had installed a Level 2 charger for his LEAF at his new home in Cambria.

Steve was more than happy to let us swing by his house and use the charger for our travels. So, last night, I had coordinated with him on the time we planned to arrive and to pick a place where he could meet us—which ended up being his house. Since we knew this would be a long day, Benswing, who also needed to recharge, and I left the group early in the

Early morning coastal fog as we depart San Luis Obispo.

morning so that we could recharge and then meet up with the group later in the day; we wanted to make the climb to Big Sur together.

We left around 8:00 a.m., and, aside from a little early-morning coastal fog again, the drive was easy to Morro Bay and then on into Cambria, where we arrived about 9:15 a.m. We hunted up Steve's house, which was a bit hidden in a wooded community, and he greeted us with open arms.

Steve showed us how he had installed a portable 220v Level 2 charger at his house so that he could recharge his LEAF. Most LEAF owners have permanent charging stations in their garages to charge their cars, but, for Steve, this was his primary charge spot because the closest public charger was in Morro Bay some twenty-two miles away, and, for shopping in SLO, it was a good thirty-five miles away. We did face a dilemma, though, as we couldn't charge both the LEAF and Ben's motorcycle at the same time with only one charger. While we charged the LEAF on 220v, we charged Ben's motorcycle on 110v. Now all we could do was wait.

After some discussion of our trip's ups and downs, Steve offered to take us into town, which was three–four miles away. Our first stop was a surprise. Steve had heard us talk about interesting places along our travels, and he took us to the Red Moose Cookie Company. Red Moose is a relatively new company, specializing in "made from scratch," high-quality cookies, whose primary ingredients are "LOVE and BUTTER." Derived from an old-fashioned style and from original recipes, the company ships custom orders not only across the country but across the globe! And, most important, the cookies have a GREAT taste. www.redmoosecookie.co

After a few minutes talking with the owner and tasting the company's product, Steve took us to the Cambria Pines Lodge for breakfast. A local and tourist retreat, the lodge has a fine restaurant utilizing herbs and veggies from its Organic Kitchen Garden.

While we had a healthy breakfast, we talked to Steve about his experience owning a LEAF, and he asked us questions about what we hoped to accomplish with the Tour. Steve had recently retired from a law firm in San Francisco, and he had moved with his wife to Cambria basically to escape the hustle and bustle of city life. Certainly in Cambria, he could do that! I suppose his owning a LEAF was one way of making a statement as well.

Breakfast with Steve at Cambria Pines Lodge.
www.cambriapineslodge.com

After breakfast, Steve showed us how close the entrance to Hearst Castle was, since I had inquired. It was only a couple of miles north of Cambria itself. Benswing and I took a look around and snapped a few pictures. I had been to the Castle a few times and thought the Tour might enjoy hearing about it, though we realistically didn't have time to see it on this day. Hearst Castle is a phenomenal display of personal wealth accumulated by print media mogul William Randolph Hearst during the early and mid-1900s. I mean, when you can line the bottom of your swimming pool with gold leaf, you have some serious money. And he did. The castle is now owned by the State of California as a state park and includes nine key tours. It's a unique museum and well worth the stop, if you have the time to stop. Unfortunately, we did not. "Next time." So after a few minutes of looking, we hopped back into Steve's vehicle and returned to his house.

When we got back, the LEAF was almost fully charged, so we switched the 220v charger over to Ben's motorcycle. We contacted the Tour and found out that they were getting close to Cambria, so we agreed they would find a place to eat and we would meet them there for lunch.

It wasn't much longer before Benswing's bike was fully charged and we said our good-byes to Steve. I thanked him multiple times and empha-sized how much of a help he had been in solving our power problem. He wished us well, and we headed over to a café in Cambria where the Tour had gathered.

The café was pretty small, and we had to sit at separate tables, but soon Susan came over to ours. She asked me if I had something I wanted to talk with her about. By this time, of course, I had calmed down, and I guess I wasn't prepared for the question right then. And I knew I didn't want to discuss anything in front of the others, so I told her that "no, there wasn't." She didn't seem comfortable with my answer, but she let it go and returned to her table. A short while later, as I was finishing my meal, I saw Susan get up from her table and leave, obviously crying and upset. I didn't know why, so as I was leaving to support BenHop on the bike, I asked Jonathan why Susan had left upset. I remember his reply: "It's like when Mom and Dad are having an argument." This took me aback because 1) I didn't know we were having an "argument" and 2) I really didn't know what Susan had or had not heard about why I was upset with the Tour. So I walked outside to find Susan and talk to her about my issues, but she wasn't to be found. Apparently she took off on the scooter to be by herself; I'm not sure. Since BenHop was already leaving, I got in the LEAF and left as well.

Sadly, this was actually as close as it came for Susan and me to have dis-cussed the issues between us. We didn't discuss it later that day, the rest of the Tour, or even afterward, when Susan stopped by my house to drop off the equipment (the spare wheel, the backup generator, and more) that was left in the supply truck at the Tour's conclusion. But I don't blame Susan

A firetruck and a few pieces of artwork from Hearst Castle
are displayed at the ticket office.
http://hearstcastle.org

*A distant shot of Hearst Castle on the hilltop
from the Visitor Center along PCH.*

or Jonathan or anyone else on the Tour, for that matter. If there's anyone to blame, it's me.

As I mentioned previously, I have a history of accumulating my frustrations and then "lashing out" when I just can't take it any longer. I may discuss the issues in the interim, but if they don't get resolved, I don't always deal well with the potential conflict. Despite the relaxation of the drive, there was a lot going on in my mind during this time with my sudden retirement (I didn't even have an income stream lined up yet with Nissan!), a new job which was now not materializing as I had hoped, plus the *daily* managing of the LEAF's power supply locations and the route, and more. While I had ongoing issues with some of the decisions being made and with Jonathan stepping in at times, the fact was that if I had issues, I should have discussed them directly with Susan. After all, I did feel it was Susan's trip and it was Susan's money we were all spending during this entire trip. I felt it was *her* trip to run as she saw fit.

My only regret to this day is that I didn't have a heart-to-heart talk with Susan during the trip. I am sure she didn't understand what was going on in my mind at the time and why I acted the way I did. I am sure that I let her down as a friend first—and that was my fault. Susan had her hands full trying to "manage" a group of eleven very different personalities during the Tour, and I'm sure the stress just got to her, too.

Despite no discussion or resolution, the Tour pressed on. We headed north on the PCH, where there were practically no towns or housing, only the golden, rolling coastal hills, an occasional farm, turquoise blue water, and brown sandy beaches. It's the way California's coast has looked for hundreds of years, and its lack of human inhabitants makes it a very serene and enjoyable drive between LA and SF. A short distance out of Cambria, we were met by some lovely bay views. It would have been nice to spend some time on the beach, but it was already mid-afternoon, and we knew we couldn't take the time to do so.

Stopping for wildlife, however, was always possible! And we did at the Piedras Blancas Elephant Seal Rookery, located about six miles north of Cambria. A rookery is a place where a large group of animals breed and dwell. And, in this case, the rookery contained a very large collection of maybe 200 elephant and California seals who regularly make this their home.

The nice thing about the observation deck here was that you were truly close enough to see the animals without disturbing their natural habitat, as well as their natural activities, which included sunbathing, playing, hunting, and mating.

The rest of the Tour caught up with us here, as did Susan. She never even took off her helmet and stayed pretty much to herself during this stop.

Everyone enjoyed the time we spent here. Sean borrowed my camera for the zoom lens and then proceeded to take plenty of "extra" pictures for me.

We then stopped for a short time at the Hearst Castle Visitors Center—a second time for Benswing and me. Again, we didn't have time to take a tour, so we moved on until we came to the Piedras Blancas Light Station.

A waterfall and a sandy beach made this bay an attraction for visitors along the coast.

Listed on the National Register of Historic Places, the light station is a reminder of the critical role lighthouses played in maritime navigation; it now serves as a wildlife sanctuary and historic park. You can get a guided tour of the light house and the grounds.

We enjoyed the view for a bit and kept pushing on. To this point, we were still in flat lands and rolling hills, so travel was easy, but the mountains lay just ahead.

A few more hills and a jog around a mountain, and we reached Ragged Point Inn and Resort. At about a third of the way from Cambria to Big Sur, this was to be the starting point for the mountain climbing portion of this leg. But before we started uphill, we took a few minutes to use the restroom and enjoy the resort amenities, which almost resembled a rest area for travelers.

There were a lot of people parked at the resort, and many of them were

The Tour stops at the Piedras Blancas Elephant Seal Rookery.
www.californiabeaches.com/beach/
piedras-blancas-elephant-seal-rookery

The observation deck got us very close to the seals and sea lions.
Sunbathing and fighting, mating or playing. I can't tell with sea lions or people sometimes.

Elephant seals and California seals comingle at Piedras Blancas.
Who can resist these faces and crooners?

not guests; that's because there was no other place to stop for miles in either direction. So passersby used the resort as a rest area, and the resort loved it.

Aside from a beautiful view along the coast, the resort also had gardens, a play area, food, and live music. We had a drink, and then were ready to hit the road.

One of the main objectives we had as we started our climb to Big Sur was to manage traffic flow around us. Up to this point, we had been able to do this fairly easily because visibility was good and we could keep a fairly steady speed of 20–25 mph with BenHop on the bike. So, with the supply truck, and either the LEAF or Jonathan's van in the rear, we slowed traffic as it came up on us and let it pass on open road.

Going up and around the mountains, however, was a different animal. Now traffic would come up on us quickly because of our slow speed, and traffic could not easily pass us because of winding curves and double yellow lines most of the way. This meant that we would have to stop at pullout areas to let traffic pass, which required coordination between the front and rear vehicles. And while there was not a lot of traffic on the PCH, it was steady, and often traveling at highway speeds.

For the most part, I thought we did a pretty good job in traffic management. On the other hand, several of the drivers coming up behind us didn't think so. Apparently, many on the road that day needed to get to see their dying fathers on their death beds before they passed away, and we were slowing them down from getting there. Despite our emergency flashers and signs, they were not happy, and they let us know by honking their horns, yelling as they passed or showing us that they were "#1" by way of their middle finger.

The lack of understanding and courtesy by California drivers was almost appalling, and also scary. When I was in the rear, I had to constantly worry about impatient drivers doing something risky or stupid to go around us before we could pull over to let them go by. It felt like an accident waiting to happen.

Up and around the mountains we went, over and over again. The scenery was spectacular, with the bright, sunny day. Meanwhile, I was watching my power supply go down, down, down. There was nothing I could do to slow it down; I would either make it or I wouldn't. If I had an ace in the hole, it was actually BenHop—his slow speed led to my nearly perfect efficiency and low use of power.

We stopped several times for traffic and to see the fantastic views. We were now into our fourth hour in the mountains, and it was starting to

Piedras Blancas Light Station is a painting just waiting to be
painted along the coast.
www.piedrasblancas.org

Susan cruises down a hill as Ben negotiates a mountain curve.

wear on all of us. We finally came into a construction area for a bridge that was on an upward slope of the mountain. Traffic was down to one lane, and the stoplights managing traffic were long. This gave us a chance to stop and relax for a few minutes.

A view up the coast from the Rugged Point Inn and Resort.

Though the sun was now starting to go down, we could see the bridge at a distance, an elegant double arch across an inlet bay. And it wasn't long before we were crossing it. We didn't really stop because we needed to set up camp.

We still had about another twenty miles to go after we crossed the Big Creek Bridge, and my power was getting very low. I was now not only worried about not making it to the campground but that it was getting dark fast and the idea of pulling out the backup generator from the supply truck in the dark was not very appealing. Once again, it was going to be really, really close.

We stopped at the top of another

A few random shots around the Ragged Point Inn and Resort.

ridge just as the sun was setting. It was a perfect sunset, and Jonathan wanted some video. He got video, and we got some iconic pictures out of it.

We pushed hard to finish the last remaining miles. Susan had arranged for us to camp again in the state park, so we knew we'd be setting up tents in the dark.

I was down to one bar of power, and I didn't know when it would disappear. All I needed from *Green Lightning* was just a few . . . more . . . miles. I had come so far I just didn't want come up a mile or two short. What a disappointment that would be! I wanted to open my door and stick my leg out and start pushing to help: "Push! Push, Rachel, push!"

Dominique and Susan share a conversation as we start our climb to Big Sur.

Then, there it was. The turnoff sign for the state park. Had I made it? I wasn't sure. It depended on how far the campground was off the main highway. We stopped at the park entrance, and I learned that it wasn't very far to the campground at all. I had done it! Once again, I'd beaten the odds and accomplished what no one else had tried. I was thrilled! It was a great feeling knowing that I had just overcome my biggest challenge of the entire trip: I had made Big Sur with a Nissan LEAF with the help of a good Samaritan in Steve Schloop. It would be all downhill from here. And not in a bad way, but in a good way!

Not so fast, "Spark-hopper." Did you know that this campground had no utilities—i.e., no water or electricity outlets? Say what? You've got to be

A shot of the riders behind me with traffic accumulating behind Jonathan's van.

Watching for impatient drivers as I try to manage traffic flow around us.

Jonathan's crew films the riders as we chug up the mountains.

Hand signals were one way to manage traffic around us, and pullouts were critical.

In some areas, we leveled out and rode around the mountains and just above the water.

kidding me!! It had happened *again*. We had not booked the full-service campground a couple of miles further up the road; we had made reservations at the primitive campground with no electric power sources for an electric parade. " Oh . . . my . . . God!"

Benswing and I drove around the campground looking for a possible RV outlet or other power outlet. There was none—anywhere. We asked around and found out there was one power source, a 110v outlet in the main restroom. Well, it wasn't ideal, but it was something. With 110v, I didn't know how much of a charge I would actually get overnight, but I knew it wouldn't be a full charge. The only good thing was that most of tomorrow would be downhill one way or another, so hopefully the charge would be enough for me to get to Carmel or Monterrey, where

The riders wait for the lights to change at the bridge construction.

Passing through the one-lane construction zone.

there would be public chargers. No choice. I went with it. I parked the LEAF just outside the door of the men's restroom and ran a long cord to the outlet inside.

Of course, that wasn't the end of it. While we were eating dinner, guess who stopped by? Yep, your friendly female park ranger. She advised me that it was against the park rules to park by the restroom; motor vehicles have to be parked at the campsite. But after I explained my situation, she told me that, as long as the vehicle was moved before the ranger made

Scenic view stops were a time to relax, reflect, and catch up on messages.

*We went through the miniscule town of Lucia and
then reached the Big Creek Bridge, one of the most
photographed locations along the PCH.*
www.californiabeaches.com/beach/big-creek-cove-beach

rounds in the morning, I would probably be okay. I thanked her for her leniency, and she went on her way. All the others apparently had extra batteries, so they didn't need my outlet, either. Fortunately, no "anti-electric" dude pulled my plug overnight, so I was able to get a decent charge off the 110v outlet. I had just enough power to get by. That's all I needed.

More important, I had conquered Big Sur from the south side! I doubt

An amazing sunset was a perfect end to a very long,
but beautiful, day driving the PCH.

anyone else has tried that with a LEAF to this day. It is very risky, but planning and a little luck got me past another obstacle. *To Sur with Love,* and I made it.

But next time we do a *To Sur with Love* visit, I'll make sure we book a campground with 220v RV outlets.

CHAPTER 45

Day 42: Big Sur to Sunset Beach State Park

"Jay'ded."

"The Little Kicking Gas Tour Does the Little Car Show."

I awoke early the next morning and packed up my tent and sleeping bag. No one else was really stirring yet, so I walked over to our campfire from last evening and sat at the picnic table to check the maps and public charging station locations for the day. I was absorbed in my task when I suddenly realized I wasn't alone.

Someone was talking to me, but it wasn't human. It was more of a squawking. I then realized it was a collection of blue jays who were moving in to ransack our camp for leftovers. These noisy nuisances had obviously done this before because they knew what they were doing. Their loud screeches were like warning signals to tell me and everyone else to leave them to their work. They located some meat scraps under the BBQ grill and proceeded to peck holes in the plastic bag and pull out tasty morsels for themselves. I swear they were members of Bluebeard's pirate clan in another life and came back as blue jays. They even woke Sean up with their antics!

I knew I needed to move the LEAF, so I walked over to the restroom and checked the charging level. To my surprise, the 110v charge was better than I thought, between 60 and 70 percent of a full charge. That wasn't bad.

Early morning at the Big Sur campground.
www.parks.ca.gov

Sean stirs in the hammock because of all
the noise from our feathered friends.

If we were almost on the down side of the mountains, I should be fine. I unplugged *Green Lightning* and moved it over to the campsite before the ranger came by; I didn't need a fine to start my day, either.

Before long, the Bens were up, and BenHop and I prepared to leave. The others would catch up. Rachel was up, however, and told me she wanted to ride with me for the day, so the three of us set out for our destination, which was Watsonville, specifically Sunset Beach, where we would camp one final night. This leg was only sixty miles on paper, though if we had to go into the city to eat or anything, it would add some miles. In between, however, we had a consumer event that Susan had gotten us an invitation

You think this wasn't coordinated? One captain and three mateys carry out the theft. I think this guy was bobbin' for marshmallow droppings.

to along the way, a car show in the Monterrey area entitled "The Little Car Show." It sounded interesting, and we would be a guest display!

BenHop, Rachel, and I headed out and back onto the Pacific Coast Highway, California SR 1. As we turned north out of the campground, it became obvious that we had just passed the peak in the mountains and would now be descending. The road straightened out and followed the Big Sur River as we continued through some inland wooded areas until we reached Andrew Molera State Park. Here the river took a direct southwest turn and dropped into the ocean, while we continued north and descended to a flat area along the beach. Unlike most California beaches, there were cattle on this beach, with a large monolith behind them sticking out of the ocean. This, we found out, was the Point Sur State Historic Park and the California Sea Otter Game Refuge.

Point Sur Light Station sits on top of the monolith, some 361 feet above the ocean. This area of the coast was a dangerous area for maritime navigation in the 1800s and the site of numerous shipwrecks. So, in 1889, Point Sur Lighthouse was built and the light lit for the first time; it has remained in continuous operation ever since. It is open to the public through docent-led tours and is on the National Register of Historic Places.

The 1889 Point Sur Lighthouse sits atop this
volcanic rock and is open to the public.
www.parks.ca.gov

Once again, we couldn't stop to smell the roses or the lighthouse; i.e., "next time." We pressed on. We continued winding our way along the coastline, finding more beautiful beaches and scenery as we did so.

While there were some uphill stretches in the mountains here, there were also plenty of downhill stretches, so I was actually generating more power than I was using. In fact, I realized I didn't need to find a public charging station at all.

The wind was very strong on this day, and particularly noticeable at our next stop, called Hurricane Point View. The name was appropriate because it felt as if a hurricane was blowing through when I got out of the car. Rachel, BenHop, and I took a look around and then asked someone to take a group picture for us. As you can see, I took off my hat so that it didn't blow right into the ocean below!

A few more miles up the road, we came to the next major scenic spot, the Bixby Creek Bridge. One of the most photographed bridges in California because of its aesthetic design, the bridge was built in 1932 for the residents of Big Sur, who were often stranded during winter months because of the impassable condition of the old road. Some 360 feet long, it stands 280 feet above the ground and is one of the highest single-span bridges in the world.

A short distance further north, we crossed a similar bridge to the Bixby called the Rocky Creek Bridge. Almost 500 feet long, it spans Rocky Creek as it flows into the ocean; here you can find a habitat for endangered

Coastline views as we follow the PCH toward Carmel.

Trying to stand at Hurricane Point View, while BenHop powers
through the wind (below).

Ben descends to the Bixby Bridge before crossing at a height of 280 feet above the ground.

southern sea otter. As we continued north, we passed more rivers flowing into the ocean, such as the Palo Colorado, Garrapata, Joshua, and Doud creeks. And more scenic spots, like Palo Colorado Canyon, Rocky Point, and Kasler Point.

A little later, we passed Garrapata Beach and Soberanes Point, before we finally saw some housing and human interface. We had reached the southern edge of Carmel. We slowly made our way through Carmel Highlands and then the city of Carmel. We passed some great scenic spots, like the Pebble Beach Golf Course, the 17-Mile Drive, Los Pinos Lighthouse, the Monterrey Aquarium, Fisherman's Wharf, and the Monterrey Cannery area, because it was now close to noon and The Little Car Show was starting in the town of Pacific Grove. As they say, "the show must go on." So we rode directly to the show; i.e., "next time."

When we arrived, the show was just getting started. We were given a pass to get in and found a parking spot along the curb, and "voila!" we were now part of the vehicles on display for passersby. About an hour later, the rest of the Tour arrived, and we had a full electric-vehicle display. The event was in its fourth year and was well attended, with a fairly large crowd. As a result, we had a decent number of curious onlookers who wanted to know about the LEAF, the Ride the Future Tour, and electric vehicles in general.

The Little Car Show is an event sponsored by Marina Motorsports, Inc., a nonprofit charitable organization featuring family-oriented, automotive events. For this event, they allowed all marques up to the first 100 fossil-fuel-powered micro, mini, and arcane vehicles under 1,601cc, and all-electric vehicles.

We spent the next three–four hours talking to folks as they passed

View of the coast from the Bixby Bridge area.

A good view of the Bixby Bridge, with its large arc spanning Bixby Creek.
www.californiabeaches.com/beach/bixby-creek-bridge-beach

Green Lightning takes a spot in line with other small cars at *The Little Car Show.*

http://marinamotorsports.org/events

by. Because this was California and particularly the Bay area, there was a lot of interest in electric vehicles. And it was fun to interact with interested people.

The show itself was intriguing, because all of the vehicles were, well, "little." There was a wide variety of types and ages, unique European brands, and more on display.

As the show concluded, we went into downtown Monterrey to do a little shopping. We found lots of unique shops and some wine-tasting venues. After that, we went to the nearby beach to check it out, but we didn't stay long.

We still had to drive to Sunset Beach and set up camp; that was another thirty miles to go. So, around 6:00 p.m., we continued our journey and left Monterrey. We passed the Fort Ord Dunes State Park and drove through a lot of flatlands.

Making good time, we reached the south Watsonville area about 7:30 p.m.; here the rest of the Tour headed west to the campground, while Rachel and I continued north into Watsonville to 1) find a level 2 public

The Little Car Show had a wide variety of brands and ages of vehicles.

charger for the LEAF and 2) find a level 2 birthday cake for Sean, since it was his birthday and we wanted to celebrate.

Rachel and I searched around town for a grocery store that might have a cake; it took a couple tries, but we found a cake with a

We checked out one of the local Monterrey beaches while in the downtown area.

silly little man fishing in a boat. We didn't know if Sean liked to fish, but, on this night, he would! We then located the Level 2 chargers, and I plugged in. We now had about two or more hours to kill doing, well, something.

We decided to walk around the area, checking out stores and restaurants. We finally found a Mexican restaurant that we liked and that was open late—we didn't go in until around 9:30 p.m.—so when the Tour was done setting up camp, they met us at the Mexican restaurant, so we could all have dinner together. We had a lot of fun that night. It was one big party in Sean's honor. In fact, we never left the restaurant until 11:00! Don't worry: we left a good tip. We arrived back at the campground at around midnight. Thankfully, Sean had set up a tent for me so I didn't have to do it. And before long, we had all crashed after another long day.

European brands were common at The Little Car Show.

Sean's birthday cake, complete with a fisherman making the catch of the day.

CHAPTER 46

Day 43: Sunset Beach State Park to Burlingame (San Francisco)

"Taking the High Road Is Mural-ly Correct."

"One Last Picture Perfect Day."

Nobody got up early Thursday morning. Everybody was exhausted from the late night before, and, more than likely, a couple of us might even have had a bit of a hangover. But then again, there really was no reason to get up early. We were knocking on the door of victory, just a few miles from San Francisco. In fact, as hard as it was to believe, tomorrow would be our last day on the Ride the Future Tour. There was certainly a sense of relief, as well as sadness, as we got ready for another day of travel.

With no refrigerated goods any longer and Watsonville about six miles away, breakfast became a collection of packaged snacks being handed out from the supply truck—this to avoid mass leftovers after tomorrow. Ben and I discussed our route for the day, where we had a number of options:

1) head directly into the cities via the busy U.S. 101 and ride in traffic to Burlingame, our endpoint today; this was the shortest route;

2) continue on California SR 1 north until it met California SR 17; then continue north into the back side of San Jose; more back roads, but still considerable city traffic;

3) continue on California SR 1 north until it met California SR 9; then continue northwest to Skyline Boulevard before snaking over to El Camino Real and into Burlingame; this route consisted mostly of back roads and had the least traffic, but it went through some hills and had crooked roads, so we didn't know exactly what to expect.

We chose option #3. Scenery over traffic *any* day.

The quickest route would be about seventy-five miles; we knew ours would be a little longer than that since we'd be taking back roads. Still, a relatively easy travel day. Benswing and the documentary crew were headed into Santa Cruz to do some filming at the Zero factory where Ben's motorcycle was built. The scooteristas were off to the Boardwalk in Santa Cruz. That meant it was just BenHop and I.

It wasn't until about 10:00 a.m. that Ben and I headed out of the campground. However, since we had arrived at dark, we had to check out the beach itself to see what it looked like because the sandy hills surrounding our exit road obscured our view. Ben and I walked down the walkway to the beach and took in the landscape. It was pretty enough, isolated in some ways. We knew that this was the last view of the beach we'd have on the Tour, because soon we'd be turning inland toward the SF mainland.

With a final sigh, we hit the road again and made our way back to the Pacific Coast Highway, our companion for some 450 miles from Laguna Beach to Santa Cruz along the California coast. The last section we would travel on the PCH went through some low-lying flatlands before we got into the city of Santa Cruz. We had some light city traffic, and then we ran into California SR 17 and jogged over to California SR 9.

California SR 9, it turned out, followed the general path of the San Lorenzo River and wound its way through a number of small and large parks, many of them wooded. As a result, there wasn't a lot of traffic to deal with, and the scenery was quite interesting. One of the parks was the Henry Cowell Redwoods State Park, where we rode past a vista point and several hiking trails and rode through a large forest of redwood trees. These majestic trees provided a natural canopy of shade to ride through; it was great to see they were being protected in the park.

One thing I wasn't expecting, however, was an elevation gain. It wasn't a real issue, but we were climbing up some sizeable hills as we drove through the park, which is why the Redwoods were plentiful. So between the shade and the elevation gain, the ride was a pretty pleasant

Walkway from the dunes down to Sunset Beach, and Ben wanted a memory picture.

one temperature-wise.

As we moved on, we came to the little town of Felton. I left Ben for a few minutes here to check out a covered bridge, since one of my passions is to hunt up old bridges. In New England, you can find a lot of them; in California, this was a rarity, in part because the weather is rarely bad enough to warrant building a covered bridge. The bridge was built in 1892 over the San Lorenzo River and is about eighty feet long. The bridge's claim to fame is that it is the tallest covered bridge in the United States, and its massive support trusses are impressive. While no longer used, it became a California Historical Landmark in 1957 and a National Historic Landmark in 1973.

I caught up with Ben, and since it was getting close to lunch, we stopped at the next "big" town, which was actually a small town called Boulder Creek. As you drive into this small community, you can't help noticing the large murals painted on the side of the local hardware store.

The murals were painted by John Ton in 2001 and depict a bit of Boulder Creek's history, which centers on the logging industry. In the

Passing an old, wooden railroad trestle as we ride through Henry Cowell State Park.
www.parks.ca.gov

The tallest covered bridge in the United States is in Felton Covered Bridge County Park.
www.scparks.com/Home/Parks/ListofAllCountyParks/
FeltonCoveredBridge.aspx

mid-1870s, the industry built the San Lorenzo Valley Logging Flume (an elevated water transport for floating logs between locations) from Boulder Creek to Felton, about seven miles away. The logs were floated to Felton and then transported by rail to Santa Cruz harbor for further transport by ship to other ports. The murals depict the large flume and later a railroad that were once at the center of the town. Very well done, the murals seem to perpetuate the town's image of being a small but rugged logging town.

We grabbed a quick bite and continued north on California SR 9. Shortly after we left town, we began a long but steady climb upward, climbing from below 500 feet in Boulder Creek to somewhere close to 2,100 feet in the Saratoga Gap Preserve. We passed a number of YMCA and Scout camps in a somewhat remote area.

I was surprised that 1) we were climbing so much in this area near San

The Boulder Creek Hardware murals illustrate the logging flume, which floated logs to Felton.

Another wall on the Boulder Creek Hardware depicts the railroad in town. www.facebook.com/ bouldercreekhardware

Francisco and 2) that this area was so remote when it was so close to a major city. Yet I suppose that's what made it special. We were so close to SF, yet it felt as if we were hundreds of miles away. This was a great place to escape from the city!

As we entered Castle Rock State Park in the Santa Cruz mountains, we found several hairpin turns, as well as scenic overlooks, which told us that we'd made the right decision on our route for the day. We had gone from a picturesque beach to picturesque mountain views in the same day,

BenHop and I enjoy the view at Castle Rock State Park.
www.parks.ca.gov

Ben Hopkins near the Saratoga Gap Preserve.

making this one last "picture perfect" day on our Tour.

When we reached the Saratoga Gap Preserve (a regional park for equestrians and hikers), we turned north onto Skyline Boulevard. Here we ran along the ridgetops through more wooded areas for a while before the road opened up and we got a great view of the San Mateo Bridge and the San Francisco Bay area.

From this point, we could practically see our final destination in Mountain View. And soon we'd be heading downhill to Burlingame.

Shortly after our great view of San Francisco, we started our descent, going down to California SR 84, which continued down past Wunderlich Park and straight into Redwood City. We were now in the city and the traffic once again. California SR 84 took us straight to California SR 82,

From Skyline Boulevard, you can see the San Mateo Bridge and San Francisco Bay.

Riding along Skyline Boulevard near Redwood City amid lots of tall redwoods.

Skyline Boulevard view of San Francisco. We had the end in sight!

which was El Camino Real, a major north-south road paralleling the busy U.S. 101. Only fifteen miles from the edge of the city, we were soon at our hotel in Burlingame, the Crowne Plaza at San Francisco International airport. The hotel was not exactly prepared for a group of electric bikes, scooters, and motorcycles in their valet parking area! Let's just say, it's not something people see every day on a very busy street in a major metro environment.

BenHop and I arrived around 3 p.m., and it wasn't long before we were all together again. We were all excited that we were so close to the Finish Line. All of our hard work was finally going to pay off just as we had planned at the start forty-four days ago. It was exciting! Afterward, we would celebrate. Tomorrow would be a big day.

We had dinner at a nearby restaurant, and then I had Sean drop me off at Sunnyvale Nissan to recharge the LEAF overnight. I hadn't driven a long distance, but the unexpected climbs in what turned out to be the Santa Cruz Mountains had taken a toll on my power. In fact, when I finally got into the city and drove to the hotel and had dinner at the restaurant, my power had gotten very low. But, by now, I was used to "Battery Level Is Low" messages; this was now just a "well-managed" power day!

CHAPTER 47

Day 44: Burlingame to The Golden Gate Bridge to Mountain View

"We Get a Little Foggy About the Route."

"A Google Search Turns Up an Exception Reception."

"A Goal Is a Dream with a Deadline."
—Napolean Hill

After forty-four long, back-to-back days, today would be the day that we met the deadline, achieved our goal, and realized the dream that Susan had created for us months ago. It was hard to believe A) that we had done it on schedule and B) that it would all be over tomorrow. But while we all had mixed emotions as we met for breakfast this Friday, August 16, 2013, our outward emotions were all excitement. There was a growing sense of pride of our accomplishment and an eagerness to celebrate it. All we had to do was execute the plan one more time.

Our final ride would bring us to the Golden Gate Bridge; then we'd take a ceremonial drive across the bridge and back, and then a ride back south to Mountain View, where Google's headquarters were located, for a triumphant crossing of the Finish Line. The total distance wasn't too bad, around seventy-five miles, but most all of it would take place in heavy city traffic through the heart of San Francisco. There was also the little issue of possibly being thrown off the bridge or arrested when the toll booth staff saw our little parade trying to cross the Golden Gate Bridge; we didn't know whether our vehicles would be allowed or not, possibly because we'd be driving too slow for traffic, but since it was officially a U.S. highway crossing—the busy U.S. 101—we assumed the worst and that we would not be welcomed. We'd literally "cross that bridge when we came to it!" and, at a minimum, Jonathan would get whatever happened on videotape.

Benswing and I had discussed the best route to take north to the Golden Gate Bridge. The fastest and most direct route was the busy U.S. 101. So that was out. There was also I-280, which became California SR 1 in Daly City, but it would likely be very busy with traffic. So we decided to go north on El Camino Real to Daly City and then go west to Skyline Drive and follow the road along the coast up to the Golden Gate Bridge. This route would have the least traffic and would also be the most scenic.

Someone took me down to Sunnyvale Nissan to pick up the LEAF, and, by the time I got back, everyone was pretty much ready to go. Since Jonathan wanted to film our travels today and I knew the route, I would lead the group down to the bridge and across it. Just before 9:00 a.m., we were off.

The first part of our trip was uneventful, just a lot of city commuter traffic along El Camino Real; then we drove west across the peninsula to Skyline Drive. It was a nice, sunny morning, and it looked like a great day for the record-breaking event. However, as we started north on Skyline Drive, things changed.

It's a fine-looking day as I lead the group out for the Golden Gate Bridge.

It turned out that Skyline Drive was just that, a road that ran on the top of a small ridge of hills along the coast, with a number of residential communities mixed in. This higher elevation, however, was apparently a nesting area for coastal fog, and we drove right into it. We initially thought that the fog would be temporary and that we would drive out of it, but that didn't happen.

The fog became problematic because

1- it was somewhat dangerous due to our slower speeds and the possibility of other drivers being unable to see the emergency flashers on the supply truck or on Jonathan's van or being unable to see the riders if the other drivers cut back into the right lane after passing Sean and Jonathan,

2- the fog would build up on my windshield and on the helmets of the riders, making visibility difficult, and

3- the fog was *really* cold, and, even with gloves and scarves, the riders were cold as well.

We stopped a couple of times so the riders could regroup and warm up a bit. We kept expecting to drive out of the fog at some point, so we continued on Skyline Drive, but into more fog.

When we stopped again, the riders told us they were freezing their tails off, so we found a place to pull off and discuss the situation. We all agreed

We encountered heavy, cold fog when we started on Skyline Drive.
Stopping to give the riders a break from
the cold wind and fog and to clear their visors.

we needed to do something different. So we decided to go back down in elevation and hope the fog wouldn't be there. We elected to go back to California SR 1 and continue north to the bridge, so, at the next opportunity, I turned east, and we came down off of the coastal hills and back into the city. Sure enough, we got out of the fog! This gave all the riders a

The Bens and I discuss going back down in elevation to get out of the fog.

chance to warm back up as we followed California SR 1 briefly through Golden Gate Park and the Richmond District of San Francisco, and finally into the Presidio area.

The Presidio took us back into some small hills as we approached the coast on the north side of the peninsula. We knew we were now getting close to the intersection of U.S. 101 and the actual bridge, so we took the last exit off California SR 1 to find a location where we could strategize our crossing. This took us into a parking area directly across from the entrance to the bridge, which *normally* would have given us a great view of the bridge. But not today. Today you could barely see it.

We huddled up at the small park nearby and discussed how we would approach our crossing. Jonathan had some questions and decided it would be best to drive across the bridge in the van so he could see what would work best for filming, as well as our exit strategy on the other side of the bridge. He took a couple of the riders with him, and they drove across the bridge and back; then he did it a second time, and I rode along with him to see what our challenges would be.

Like our previous large-bridge crossings, the tricky part was our initial entry because of the fast traffic and the short ramp. I would wait until we

The great structure of the Gold Gate Bridge is barely visible in the fog as we prepare to cross.

Our initial entry onto the Gold Gate Bridge was foggy, and a bit scary, with heavy traffic and winds.

had enough clearance for the whole Tour before I headed out. Jonathan, in wanting to get great video, would pull to the left when he could, and Sean would block traffic from the rear with the the supply truck's emergency flashers going. The crossing would be hairy, especially for the riders with the crosswinds on the bridge, but at least the traffic wouldn't be doing quite the speed we had encountered in Memphis. There were also Homeland Security vehicles stationed on our entry and on our exit end of the bridge, and we speculated on whether they would stop us or not if they saw us.

While driving over the bridge, I felt exhilarated. Aside from my concerns about traffic, I was thrilled to be crossing such an iconic structure and location as part of our final act. For Benswing and BenHop and the scooter riders who were braving the cold, fog, and winds on the bridge, I'm sure the crossing was a bit scary but equally thrilling. We were making history—and why not under adverse conditions just as the trip had been? Oh, and the thrill and the threat of potentially being arrested was also in there somewhere, too!

Nevertheless, the ride across was rather uneventful. Simply put, it went as smoothly as we could have hoped. And, surprisingly, no police followed us as we exited. (I'm still

It was a little less foggy as we made our crossing, but it was not foggy for the entire distance.

BenHop leaves more black tire marks on my bumper as I break wind for him across the bridge. Evan hangs outside the van to capture our crossing in the fog on the Golden Gate Bridge.

not sure why not because Homeland Security must have had cameras everywhere for terrorism threats. Maybe we just got lucky, and the police were making a Krispy Kreme run at that time. Or maybe they were camera shy.) Regardless, we completed our crossing exactly to plan.

We pulled into our little parking lot park area just off the exit. We thought we were done, but Jonathan wanted more film and angles, so we did a repeat performance. This one was even smoother and without incident. I did note that there were cameras taking license plate pictures since there were no toll booth collectors, which meant if you didn't have a California toll transponder, you would get a ticket. I would get two. Sorry, Nissan!

As we raised our fists in victory and crossed the Golden Gate Bridge the final time, we knew our trip was truly complete. It was a glorious feeling, but we'd gone as far as we would go. The only thing left now was to cross our Finish Line at Google Headquarters in Mountain View, even if we had to crash the gates to make our presence known!

Susan had been in contact with Tiffany regarding Google a few times, but it was hard since she was traveling. Google's Public Relations Department had turned us down because, on the following weekend, it had another electric-car event that it sponsored, and the event was on its premises. Thus, it apparently didn't want to cause confusion with the press or the participants, which might happen if Google sponsored or supported our electric-parade event. Google's stance was understandable, but certainly disappointing.

Susan hadn't given up because she had always envisioned the Ride the Future Tour being welcomed by Google as an example of the viability of electric vehicles, which the company certainly supported. So Susan got Tiffany, who had helped us set up a number of consumer and press events, to help out. And Tiffany was a bulldog when she set her mind on something. She set her sights on Google and started making calls all around the company to find an executive who would override the Public Relations Department's rejection of us to have our Tour Finish Line and final party at the company's location. After dozens and dozens of e-mails (you can't easily call in), she got a response. One executive thought our plan would be a good idea. The executive's thought wasn't ideal in that Google didn't welcome us with open arms with corporate executives or press, but it did allow us to have a Finish Line Celebration and to tell its employees that we were arriving if they wanted to visit with us. Google's decision was something. And we had a venue for our final event. Well, not quite. It seems the area inside the gates is a park owned by the city. That meant Tiffany had to get a permit from city officials *the morning of our arrival*! She received it just before we arrived. Thank God. Now we could call Google a "goal" and not a "detour." Kudos again to Ms. Tiffany and to that Google executive— in Maintenance, was it? (I can't recall.)

When we completed the second crossing of the Golden Gate Bridge, we again returned to our little, triangle-shaped parking lot at the intersection of Lincoln Boulevard and Merchant Road, just off the first bridge exit. Here we stopped to relax and talk and discuss the logistics of our

We set up the electric vehicles side by side for Benswing to explain charging and cost differences.

arrival at Google Headquarters. Jonathan wanted to make sure he had the right footage for the end of the documentary, so he wanted us to gather for a final parade into Google's campus. He also wanted to do some quick filming right there in the park to explain the differences in costs, batteries, and charging requirements between the LEAF, the motorcycle, and the scooters, so we set up a quick display of the vehicles, and Benswing did a great job of explaining the differences on camera. This clip would be used in one of the early sections of the documentary so everyone could understand the basic challenges we faced as our trip unfolded.

Off we went toward Google, back down California SR 1 through the Presidio area and through the center of the city, then crossing over to El Camino Real, which we could take all the way south to Mountain View. We arrived at a restaurant which we used for a staging area at about 2:00 p.m. and then had lunch. We were very close to Google headquarters and would approach on a low-traffic street, which would enable Jonathan to film everything easily. He decided the order for our entry: I in front, followed by Benswing, BenHop, and then the scooteristas and Sean. He got set for his filming, and we started for the entrance. However, just as we got to the entrance, we were told to stop. Jonathan didn't like the first take and had us make a couple of adjustments, and we did "Take 2." This time, Jonathan was good with it, and I led the group up the drive of Google's main campus.

The Ride the Future Tour members at the Golden Gate
Bridge terminus of our trip.

Only Jonathan and Susan knew what to expect as we drove in. We were surprised to see a Finish Line banner strung across the entryway and a small group of enthusiasts there to greet us as we drove and rode in. We came in with horns blowing, all of us screaming and yelling, our fists raised, and more, for ourselves—and for the camera. There were cheers from the crowd and from a few selected individuals ready to bombard us with Silly String as we came across the Finish Line. The crowd wasn't huge, but it seemed big as we drove up! We then parked the car and vehicles to celebrate.

The crowd was a collection of electric-vehicle enthusiasts, some local LEAF-owner group members that we had met the night before at a restaurant, friends, Ben Hopkins's A2B electric bike sponsors (they had a small tent display), some employees from Google who were intrigued, and a couple of press folks. Oh, and one special attendee. Stuart, who you will remember had left the Tour back in Oklahoma, was there to be a part of the celebration with his son Sean and The Tour as well! It was great to see Stuart again, and all of the attendees, for that matter.

After a couple of hours, the Google folks headed home, and the party was over. Thus, the air was now out of the balloon. Soon all of us would be going our separate ways, and the Tour would be no more. It was a sad period of time. And not even dinner with my Tour mates and our EV friends nor drinks at the English Pub we found later could make up for that empty feeling. I'd made some lifelong friends during the Tour. I got a rare opportunity to experience what it feels like to be a celebrity—albeit a small-time one, and it was fun! While we were celebrating our achievement, for me at least, the time was bittersweet.

And celebrate we should! The Ride the Future Tour had done what it set out to do: to show that electric vehicles were viable today and to

Susan serves up a champagne toast to everyone's delight.
Jonathan and his crew capture all of the celebration action.

further educate the public on their benefits. The Tour wasn't perfect; we had a learned a lot about what we should have done. The Tour wasn't always pretty. It didn't get a lot of recognition. But it did accomplish what no other group had ever done: it had crossed the United States in all-electric vehicles. Even if we didn't get credit for setting any world records, we could still claim that, with a good plan and a flexible cast of characters, we had accomplished something unique. Something we could all be proud of, even if few people knew about it.

The Ride the Future Tour, our little "Kicking Gas Tour That Could" proved that, over 4,400 miles and over all types of terrain and obstacles, *electricity* is viable as a power source and *can* replace *gasoline* engines.

Now all we have to do is "Take Charge!" and make electric vehicles a part of mainstream America.

The Ride the Future Tour and "Friends" at the Finish Line.

EPILOGUE

The next day, we all started going our separate ways. As sad as it was that the Tour was over, we said our good-byes, and then it was back to our individual lives. No more crowds. No more press. No more cameras. And no more Tour.

I spent the weekend in San Francisco relaxing, visiting A2B's office, and seeing a few sights. I then drove back down to Los Angeles to drop off the LEAF at Nissan—but I took the fast route this time, via U.S. 101. Of course, I had to recharge a few times during the day, but I had no time pressures; I simply enjoyed the California sunshine along the way.

I turned over Green Lightning to the Nissan Regional Office in the LA area. No one greeted me or came to see the car that would later be confirmed as having set a Guinness World Record. Instead, Green Lightning was likely disrobed of its unique wrap and put out to pasture like an old mare to provide simple commuting pleasure for a Nissan employee and later a retail buyer. It deserved a better fate; it deserved to be put into the Nissan museum of collected vehicles that illustrate the evolution of a vehicle over time. If electric vehicles are to be a key part of our future transportation in the United States and the world, then Green Lightning blazed the path to show what could be done! But that "better fate" didn't happen. Why? You'll have to ask Nissan that question.

Speaking of the Guinness World Record, after a whole lot of documentation submission including

- a daily journal of our travels with dates, times, locations, activities, and more;
- a copy of the electronic GPS tracking of our locations and mileage;
- detailed examples of the planning documents for routes and vehicle charging;
- signed witness books at our locations along the route;
- photos, press articles, vehicle inspection documents, and more;
- and Guinness application documents,

I was awarded a Guinness World Record certificate for the "longest distance in an all-electric (non-solar) vehicle" at 3,534.77 miles for driving from Charleston, South Carolina, to Laguna Hills, California. It's framed and hangs on the wall in my house.

If you look in the 2015 edition of the Guinness World Records book, I'm on page 156. Thanks to some local publications, my neighbors know about it, as well as a few friends. That's about it. But the GWR is mine and not too many people have one.

You may have noted that my record was for 3,500 miles and not 4,400. Guinness can be unbelievably challenging on validation. They cited a

My Guinness World Record hangs on the wall,
and I'm shown on page 156 in the 2015 GWR book.

My personal memento: a T-shirt with all the cities signed by all of the Ride the Future Tour members—with their "best" wishes!

problem with my documentation on the GPS I used (a problem which I still can't explain) about the last part of my trip from LA to SF. So Guinness would only give me credit for Charleston to LA. I had to take it.

But the bigger point here is that mileage documentation was the sole reason why Ben Hopkins and Susan were denied GWRs, even though they completed the Tour just as I did. This is such a tragedy. I mean, can you imagine riding a bike or a scooter 100 miles a day for forty-four straight days in all kinds of weather, doing something no one has done before, setting a long-distance record, and not getting credit for it? Ben and Susan deserved a better fate for their efforts, and if you get the chance, maybe you can send them a "congratulatory salute" via their Facebook pages. I'm sure they would appreciate it.

There were a lot of "so close" examples surrounding the Ride the Future Tour, aside from the denial of GWRs for Susan and Ben. For example, I had discussions with AeroVironment to do a series of publicity articles touting its new "Turbocord" as the key to my achievement of a Guinness World Record, yet, at the last minute, AeroVironment pulled the plug and never said why. Nissan's never getting on board to sponsor the event was another "so close" that would have made a world of difference in the PR and consumer exposure. Funding to edit and to market the documentary was another "so close" example. If we could do the Tour all over again, there are many things we would do better.

It's easy to get hung up on these "so close" disappointments, but I prefer to recognize what the Ride the Future Tour did accomplish. What the Tour did accomplish was to prove that electric vehicles are a viable alternative to fossil-fuel vehicles today. Whether we realized it or not at the time, that was our primary goal, and, as a team, we accomplished that goal.

We accomplished this viability goal in three main ways:

1) Proving that electric vehicles were reliable and could travel long distances without issue today.

2) Showing that, with a developed infrastructure, power and distance are not limitations for electric vehicles:
   - *Electric power* is *everywhere today,* be it 110v, 220v, or even 440v to power electric vehicles. It may be hiding in hotel or motel air conditioners, vending machines, parking-garage ceiling outlets, and more, but electric is there if we simply access it.
   And if you have power, especially one that's not an expensive fuel, then distance is not an issue, either.
   - The infrastructure in 2013 was insufficient in terms of the number of chargers available and in terms of their operable condition. This insufficiency is something we *can* resolve easily enough. The infrastructure is better than it was in 2013. And it will get a big shot in the arm as a result of Volkswagen's "Dieselgate" settlement with the U.S. Government for emissions errors. Volkswagen has formed a company called Electrify America which is *required* to spend $2Billion to build an infrastructure for electric vehicles, including over 3000 vehicle chargers over the next 10 years. And if they ask me, I'll tell them where a few of them should go!

3) Expanding public interest in electric vehicles.
   - The Tour met a fair number of consumers; they see the benefits of electric vehicles and want to explore the option. Electric vehicles are even starting to become fashionable now.
   - Consumers told us they are willing to make small sacrifices for the benefit of improving the environment *and* potentially to lower their costs.
   - Consumers also see the tie between electric vehicles and sustainability, though some don't know exactly what it is.

These are the elements of the *Tour* of which I am most proud; they are many of the main reasons why I did the *Tour*. I want to leave this *Earth* in the same beautiful condition that I got to enjoy it, for my kids, and for my grandkids.

My colleagues on the *Tour* and I believe that all the oil that we are pumping out of the ground is not perpetual and will run out at some point. If we don't develop alternatives and the oil does run out, the results could be catastrophic.

Finding proactive alternate solutions should be a priority for the world, and electricity is one alternative that's viable *today*. The United States should lead the way to show the world how to make electric vehicles a part of everyday life. But it is countries like Norway that are actually leading the way. Norway has a fleet of plug-in electric vehicles which is the largest per capita in the world, with Oslo recognized as the EV capital of the world. The market concentration is 14 times *higher* than the U.S. (the world's largest market). *And*, 98% of the electricity generated in the Norway comes from hydropower—not coal burning! Government incentives were used to target an initial goal of 50,000 electric vehicles which included exemptions from vehicle fees, purchase taxes, and 25% import taxes, etc which made electric car purchase prices competitive with conventional cars. Local authorities were also granted the right to decide whether electric cars can park for free and use public transport lanes. Light-duty plug-in electric vehicles registered in Norway totaled more than 135,000 units at the end of December 2016, making the country the largest European stock of light-duty plug-in vehicles, and the fourth largest in the world after China, the U.S. and Japan. Norway, I salute you!

My point? It can be done.

Speaking of my colleagues, while I've crossed paths with a couple of members since the Tour ended, without a formal reunion, it will be tough for all of us to meet, because we're spread around the globe:

➤ Stuart and Sean headed back to Hawaii and school.
- Stuart found a wife in Cambodia while working there and is now retired in Hawaii. I wonder if he still wears a Speedo.
- Sean is a research assistant for the School of Public Health at Harvard. I confirmed he hasn't driven into any buildings since our Tour.

➤ Ben Hopkins met with the A2B Electric Bicycle Company to give them a download of his likes and dislikes about the product and then returned to Thailand.

- Ben now works as a hotel review writer and cartoonist, traveling to various locations around the world. He recently sailed a narrow boat he had remodeled and named The Blue Otter back to England. I "*otter*" call him to find out how he is, but I don't think he has a phone, or can find it. We do keep in touch a bit through Facebook and e-mail.

➤ Rachel jumped in with Terry and traveled around California on Terry's bike to various events.

- Rachel works for UNICEF in Cambodia; she travels a lot and has many international friends, and a few dog friends waiting for her back in Oklahoma.
- Terry continues to be an advocate for electric vehicles and rides to rallies on the West Coast with his dog, Charger. He recently got into a road rage argument with a heavy-duty truck owner, who thought Charger should be fetching a newspaper and not riding on a motorcycle. Go, Charger!

➤ Ben Rich returned to New Jersey to prep for the new school year.

- Ben continues to teach physics to high schoolers and to be an advocate for electric vehicles; he's made a couple of long trips around the country on his motorcycle and stopped by Nashville to see me. He's still a big "swinger," just not my type (of dance).

➤ Susan and Dominique had the luxury of driving the supply truck back across the country to Nashville and delivering all of the various parts, clothes, equipment, etc. to their rightful owners.

- Susan returned to Thailand, started an electric-scooter tour service and now does sailing excursions with her beau. I'm sure there's a new cause or dream over there somewhere.
- Dominique continues to develop her singing career and is now touring as a backup singer with a band out of California; if that doesn't work out, will the Luna Belles reunite?

➤ As for me, I've considered postretirement work but decided to travel, cruise with my daughters, launch a travel-guide website (funexcursionsinabox.com), and write this book instead. No, I don't drive a LEAF because I drive long distances to see family and have no room in my three-car garage for a second vehicle. True. I do drive a hybrid car though; more on that in the next chapter.

We all keep in touch occasionally through Facebook, which is nice. Posting an occasional picture from the Tour or wishing a "Happy Birthday" brings back memories and comments we all can relate to.

Sadly, I haven't kept in touch with Jonathan, Evan, and George. I know Jonathan made a *great* Make-A-Wish Godzilla movie for a Wish child. He is living in Chicago now and works for the Leo Burnette ad agency there. And since Shaquille O'Neil is still trying to kill foot fungus with Gold Bond, I'm sure George is gainfully employed periodically.

As for the documentary itself, it hasn't had the impact that everyone had hoped for, primarily because it never got off the ground. I don't know specifics, but funding became an issue. And I know that the initial attempt to cut it all down to a ninety-minute movie was very difficult. After all, they filmed hundreds of hours of film on us. It's still out there. If you know anyone that would love to pick up this story in actual video, *it's waiting!*

Finally, I would like to acknowledge the other members of the Ride the Future Tour and the documentary team. This trip was a very positive, once-in-a-lifetime experience for me, mainly because of all your individual personalities. The Tour did what it set out to do—to prove that electric vehicles could be used today for transportation and thus open the window a little more on the sustainability movement. I am glad I was a part of the Tour, and I hope you are as proud of our accomplishments as I am.

Thank you for all your help and your friendship.

I must add a note of thanks to Susan. It was her vision, her persistence, her patience, and even her money that made this Tour a reality. Rarely will people truly put their money where their mouths are, but Susan did. The Tour wasn't perfect—and not always fun. But Susan made it happen. For that, I will be forever grateful.

The Ride the Future Tour was one hell of a Tour! It was an adventure. It was a challenge. It was celebrity atmosphere that we all dream of having some day in our lives. But it was also disappointment—of what could have been. The world is "kicking gas" today, but will it convert to electric? I don't know. But if I'm asked to prove electric is viable again by doing another Tour, I know what my answer will be: "In an electric heartbeat."

Kicking Gas and Taking Charge!
(in a hybrid car)

CHAPTER 48

Kicking Gas Your Way Every Day

"Sustainable Is Maintainable If Habits Are Retainable."

"I Have Seen the Light! And It Is Cheaper."

I'm not a tree hugger. I'm really not. In fact, one of the concerns I had when I joined the Ride the Future Tour was that I would be the gas-guzzling, corporate outsider that forgot to turn off the lights while traveling with a group of chain-me-to-a-tree conservationists. Thankfully, that didn't turn out to be the case.

However, through the Tour, I've come to know wonderful individuals who are committed to all things eco-friendly. And while we had our differences at times, what I found common among all of us was a goal for a sustainable future. Before this trip, I had a typical, everyday awareness of what sustainability meant, but I certainly wasn't going to explain the subject to others. Sustainability is different for everyone. Truth be known, probably few of you can actually give me a definition. So this is where I choose to start the final portion of the book—a section dedicated to offering up a few reasons why you should care about sustainability and some simple, easy ideas on how you can "Take Charge" and make a difference in the world yourself.

I. My Definition of Sustainability

"'Sustainability'" is action taken to meet the needs of present society without compromising the ability of future generations to meet their own needs."

Want a more direct definition? Okay, try this:

"Achieving 'sustainability' will enable the Earth to continue supporting human life."

You probably won't have to be concerned about definition #2, but your future relatives (e.g., your grandchildren or their children) most likely will.

Why do I make such a scary "Mad Max" movie-type statement? Well, I'm an engineer by degree, and we engineers always look to numbers to answer tough questions. And, in a minute, we'll take a quick look at the numbers published by the Environmental Protection Agency (EPA) and other sources.

But, first, there is a little more clarity I would like to provide relating to the definition and study of "sustainability." Sustainability includes three major pillars, namely social, economic and environmental elements.

Each of these elements interacts with the other. (See Figure 48-1.) To use one extremely simple example, suppose you have always wanted a jet ski for the lake. You buy one to have fun with your friends (social/economic) and use it, but, in the course of playing, you spill gallons of fuel into the lake. The lake is now polluted and not in its natural state (environmental). To return it to its natural state and enable everyone else to swim in it again, money has to be spent to clean the spill up that otherwise could have been used for another social or environmental enhancement (economic). Laws are then put in place to reduce your capability of spilling fuel (social), and incentives are introduced (economic) to encourage the next jet ski buyer to buy an electric model to avert such disasters. And so it goes.

Figure 48-1: The Pillars of Sustainability

➤ Social: all things related to PEOPLE and their quality of life
➤ Environmental: all things related to the PLANET we live on
➤ Economic: all things related to PROFITS and money spent for life

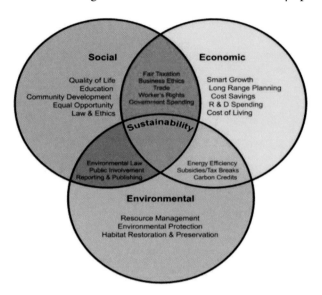

In the schematic shown, you can see these interactions and how sustainability is achieved. Effectively, it is achieved when all three of these elements (social development, economic development, and environmental protection) are in balance.

My interpretation of sustainability being the point of balance between the pillars:

We will achieve sustainability (balance) if, on an ongoing basis, we are improving our quality of life as a human race without sacrificing all of the natural resources of the planet—thus leaving the Earth habitable for future generations.

Okay, so much for textbook definitions and the macro-view; why then should you care about sustainability? Let's look at the numbers.

II. Sustainability by the Numbers

While there are a number of variables that affect sustainability, I'm going to try to keep things simple by focusing only on a couple of primary indicators in the case for supporting sustainability efforts in the world today: world population and human consumption and waste.

Figure 48-2: World Population and Growth Rate Trends and Projections

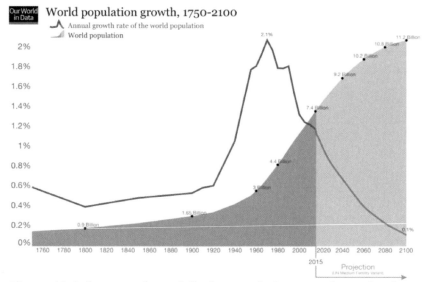

Figure 48-2 shows us that, while the population growth rate has been slowing since 1962, the sheer volume of people is forecast to continue to increase dramatically. In fact, over the last 50 years (1965–2015), the world population roughly doubled from 3.2 billion people to about 6.6 billion; in the next 50 years, it will increase 4 more billion to 10.5 billion. That's a 60% increase!

Takeaway: That's a lot more mouths to feed, as well as inhabitants who will consume and dispose of waste! Yet the inhabitable area and natural resources of our Earth will not grow.

Let's now compare populations to human consumption and waste disposal.

Human Consumption and Waste

For simplicity, I'm going to focus only on energy consumption because it is the primary driver of natural resource utilization and waste. If we look at the three most populous regions of the world and compare people to energy use, we see some startling statistics:

Figure 48-3: Population vs. Energy Consumption and GDP by Largest Regions

	% of World*		
Region	**Population**	**Energy Consumption**	**Gross Domestic Product (GDP)**
U.S.	< 5%	18%	16%
Europe	7%	16%	17%
China	19%	20%	17%

*University of Michigan Center for Sustainable Systems website and U.S. Energy Information Administration data

Takeaway: While we Americans live a great lifestyle, it's important to realize that we are less than 5% of the world's population, yet we use 18% of all of the energy consumed!

Figure 48-4: Historical Energy Consumption

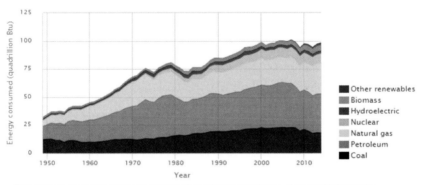

This graph does not account for net imports of electricity and coal coke, which would change the total by −0.08% to 0.24%, depending on the year.

Since World War II, the United States has realized a significant growth in energy consumption, which includes the raw materials such as oil, natural gas, coal, nuclear power, and more. This high level of fossil fuel use also means we are the largest disposers of waste in the form of solid waste materials and pollutants. China and the United States are, by far and away, the largest contributors to the CO2 elements that most scientists believe are creating the ozone layer and greenhouse effect of global warming.

Takeaway: Whether or not you believe in global warming, common sense would suggest that continued record levels of usage and growth in demand globally for fossil fuels will eventually lead to an exhaustion of a finite amount of a natural resource.

How long we have before that exhaustion happens is a subject of much debate. For example, a *Christian Science Monitor* report in July 2014 noted that British Petroleum's 2013 annual report estimated global oil reserves to be about 1.66 trillion barrels of oil and that reserves would run out in 53 years! Yet, in their 2016 annual report, the BP estimate was placed at 2.6 trillion barrels under today's technology (e.g., fracking); so effectively they found 1 trillion barrels of oil in 3 years. That's pretty amazing don't you think? More importantly, is it realistic?

My point here is that no one really knows how long oil and natural gas reserves will last. Not even Donald Trump. But what *everyone* does agree on is that demand for oil and gas through 2030 will continue to grow and that oil and gas will continue to represent at least 50% of all energy consumed—and that *assumes* a large growth in alternative fuel sources.

Takeaway: When you consider that it takes *tens of millions to hundreds of millions of YEARS* for petroleum to form under tremendous pressure by the Earth, the fact that we don't know how long these oil reserves will last suggests real risk to my kids and certainly to my grandkids. At some point, all oil will be depleted unless we start being more efficient with fossil fuel usage and find more alternative fuels like electric, solar, or wind.

How can we be more efficient with the use of fossil fuels? Glad you asked. If we look at Municipal Solid Waste (MSW) for the United States as our key indicator for efficiency of fossil fuel use, we find a dramatic growth in MSW since 1960, just as we found for energy consumption in the United States.

The total amount of MSW generated in the United States increased steadily from 88 million tons (MT) in 1960 to a peak of 256 MT in 2006. MSW generation was relatively flat from 2006 through 2013.

Only 6% of all MSW generated in 1960 was recovered through recycling, while the remaining 94% was landfilled or disposed of using other methods. In 2013, 25% of MSW was recycled, 9% was composted, 13% was combusted with energy recovery, and 53% was landfilled or disposed of using other methods.

Figure 48-5: U.S. Municipal Solid Waste (MSW) Volume Generated and Managed 1960–2013

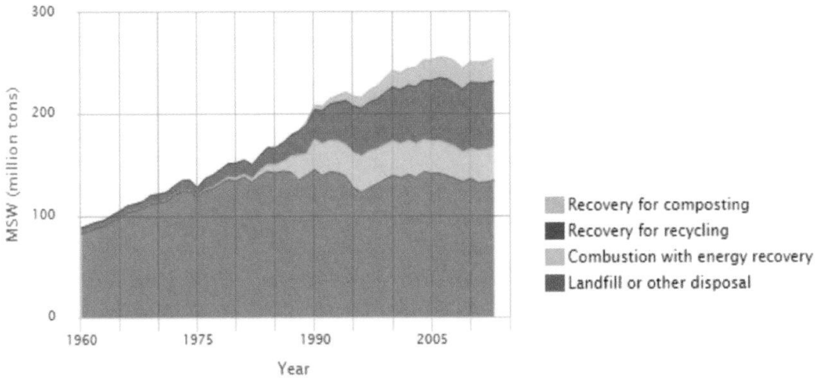

No composting data are available prior to 1988. "Landfill or other disposal" includes combustion without energy recovery.

From 1960 to 2013, total MSW generation in the United States increased by 188% while the U.S. population increased by only 75%.

Takeaway: While recycling and combustion processes have improved U.S. waste efficiency, over 50% of all waste is still sent to landfills or another disposal method. In addition, as the U.S. economy grew stronger, waste per individual accelerated just as energy usage did over the same period. The two are closely tied.

U.S. Greenhouse Gas Emissions Volume 1990–2013

When we turn our attention to fossil fuel gaseous waste, despite recent legislative and social pressures, carbon dioxide levels are 7% higher than they were in 1990, while methane was actually down 15% because of landfill and coal mine reductions. Electricity generation (power plants) and transportation are the two largest producers of greenhouse gas emissions at 32% and 26%, respectively, of all gaseous emissions.

Takeaway: with less than 5% of the world's population, the U.S. currently accounts for nearly 20% of total global emissions of carbon dioxide, methane, nitrous oxide, hydrofluorocarbons, perfluorocarbons, and sulfur hexafluoride.

WHY SHOULD YOU CARE?

While I have *barely* scratched the surface on all of the indicators of current sustainability status today, I'm going to spare you all the detail and make my summarized case as follows:

From the above data, we know

- World population is growing by 4 billion people in near-term years.

- Energy demand is expected to continue to rise because of improved economies in third-word countries, with the United States, Europe, and China remaining very strong users.

- Europe and the United States utilize a disproportionate amount of energy and fossil fuels relative to their percentage of global population and thus provide a disproportionate amount of the global waste/pollutants.

- While there has been some improvement in water and air quality and natural resource availability (timber for example), in general, the Earth's natural resources are fixed (i.e., cannot be easily replicated or created).

Takeaway: Assuming that the Earth has a fixed amount of natural resources, then these negative *growth* pressures will affect the Planet pillar and *must be offset by improved efficiencies* within the People and Economic pillars for us to achieve sustainability. We need to find activities that will reduce energy use as well as find alternative energy sources for future generations to enjoy the Earth as we know it and to survive.

As the places like Southface Energy, Furman University, Heifer International, and the Las Vegas Cyclery showed us during the Tour, the goal to be totally energy and resource independent *is* possible. But, for most of us, just moving in that direction is a practical approach to supporting sustainability. And if we can get everyone to move just a little in that direction, the synergistic effect will have a huge impact in making sustainability possible globally. Perhaps the epiphany of my trip on the Tour was that I now believe

**WE NEED TO SHOW LEADERSHIP
TO THE WORLD IN THIS ARENA!**

Unfortunately, other countries are taking that leadership role right now. For example, in the automotive world, as I outlined earlier, Norway's government set aggressive targets for electric-vehicle sales and supplemented the goals with financial and special privileges; the program has been very successful. And Geely, a Chinese company that owns the Volvo brand now, recently announced that it is converting its entire vehicle lineup to hybrid and fully electric by the year 2019, the first automotive brand to commit to a non-gas powertrain lineup.

With or without Donald Trump's support, *we owe it to the world as the largest fossil fuel user and to our children's futures to show that we care about more than self-indulgence.* This doesn't mean that we should be ashamed of our lifestyle, but I do think that we should acknowledge and take responsibility for our impact on the Earth/Planet pillar. To their credit, the younger generation seems to be in tune with this concept, and I encourage them to continue to be involved in planning sustainability activities.

III. How Can You "Take Charge" in Sustainability?

So come on America. We *can* do this! We CAN be the world leader that we should be.

A. Let's start with a few interesting sound bites of data:

➤ The EPA estimates 75% of American waste is recyclable but we only recycle about 30%.

➤ In a lifetime, the average American will throw away 600 times his/her adult weight in garbage, which translates into 90,000 pounds of trash for his/her children to deal with.

➤ 99% of lead-acid batteries and 90% of corrugated cardboard is recycled today, but only 30% of bottles and 40% of tires and electronics are recycled. Glass bottles take 4,000 years to decompose.

➤ Aluminum cans are the #1 recycled item, and thus they make up <1% of U.S. waste; in contrast, 70% of all plastic waste ends up in a landfill.

➤ You can make 20 new cans from recycled material with the same energy that it takes to make 1 new one from raw materials.

➤ 2,000 pounds of recycled paper can help save 17 trees, 350 gallons of oil, and a large amount of landfill space.

➤ Globally, *1 million new plastic bags are used every minute* at a cost of 2.2 billion gallons of oil every year!

➤ A 100 watt incandescent light bulb gives off 1600 lumens, costs about $12/year to provide illumination and will last about 1 year. A comparable LED bulb costs about $2.40/year and will last 15–20 years!

➤ Adding insulation to your homes can pay for itself within a few years.

➤ Solar power represents only 1% of all global energy sourcing.

➤ Electric vehicles are a highly efficient mode of transportation. Up to 80% of the energy in the battery is transferred directly to power the car, compared with only 14–26% of the energy from gasoline-powered vehicles.

➤ Motor oil can be recycled, but one quart can contaminate 2 million gallons of fresh water.

B. ACTIONS You Can Take That Will Contribute to Sustainability

Simple and easy. That was my focus from the list of things I found for sustainability ideas. Thus, I will only elaborate by section below and not by individual ideas—most are self-explanatory.

<u>Transportation</u>

"The time is right for electric cars—in fact, the time is critical."
—CARLOS GHOSN
CEO, Renault, Nissan, and Mitsubishi

Public	Take more mass transit when possible, or carpool.
Personal	Improve fuel economy with newer vehicles. Try hybrid or electric cars, cycles, scooters, or bikes.

This is one of the areas you can have the biggest impact on sustainability and why I want to comment. Perhaps it will change your perspective.

I am *proof* that electric vehicles are simple to use, safe, and reliable. I had minimal knowledge of the all-electric LEAF before this trip, and I ended up setting a Guinness World Record. You *do* need to make changes in your lifestyle, using an electric car as the family commuter car and planning where to recharge if you travel a longer distance, but *it can be done!* Used electric vehicles are very reasonable in cost. Perhaps good college student cars?

In addition, more and more locations are installing electric-vehicle charging stations. The charging times are coming down, especially with the CHAdeMO quick chargers. The distance that electrics can go on a single charge is increasing. So the "inconvenience" for electric VS gas vehicle driving is decreasing. And perhaps most importantly, in late 2017 General Motors announced its intent to move to an all electric vehicle line-up in the near future. These are all positive factors to support an argument for electric vehicles.

But let's make the financial case for electric vehicles, too. My Nissan LEAF had a 24 Kilowatt Hour battery. And to fully charge the LEAF in Tennessee (my home), the electricity cost is 10 cents per kwh. For a normal internal combustion engine vehicle, we will assume that gasoline price is cheap, at $3 per gallon and that the car averages 25 miles per gallon. To compare, let's assume we will drive both vehicles the average distance that the LEAF can travel on a charge, which is 100 miles (depending on topography and speed), so the gasoline vehicle will use 4 gallons to go 100 miles. Here now is a cost comparison of the two:

	Nissan LEAF	**Gasoline Vehicle**
Cost to travel 100 miles	24 kwh x $.10/kwh = $2.40	4 gal x $3/gal = $12

So excluding oil, antifreeze, and more, the ongoing cost difference between the two types is almost $10 per 100 miles driven. If you drive 12,000 miles/year, the LEAF would save you $1,200/year!

Even if you don't go all the way to electric, there are eco-friendly alternatives. I don't drive a LEAF today. I travel too far to see my daughters and parents for it to be practical. But I do have an Infiniti Q50 Hybrid—which averages almost 34 MPG! Fuel economy is getting better, and every reduction in the amount of gas used is another step.

If you've never ridden an electric scooter or bike, try one. They are *fun!* I particularly think there is an opportunity in the electric bike market; it makes the rides uphill *much* easier. Use them for fun or for local errand runs, to visit neighbors, or just to get out with nature in a local park.

Waste

Recycling	Start a community effort, or maximize your utilization.
	Reduce use of plastic bags.
	Recycle all oil, batteries, ink cartridges, paper, and more.
Compost	Utilize food and lawn clippings to fertilize gardens and soil.
Water	Install lawn rain-sensing sprinkler controls.
	Reduce water time for 1) taking showers and 2) for brushing teeth.
	Install water-saving toilets, sinks, and more.
Food	Fix smaller meals to avoid spoilage and also freeze leftovers.

There is another *huge* opportunity for improvement. As I showed in the previous Numbers section, recycling can *dramatically* reduce the energy needed to make new cans and bottles. Yet many communities do not even have a recycling program.

When I lived in California, we had separate recycling bins for different types of recyclables. It was organized and worked great. In Tennessee, all my recyclables go into one bag, and all my grass clippings and twigs go into paper bags; it's new here, but it works. My parents, on the other hand, still live in the small town I grew up in, and, during the week, there are at least four–five different refuse companies that go through the community to pick up trash. Yet the city does not mandate a recycling program for any of them. So where does all that trash go? Yep, to a landfill. This is why the United States recycles only about 30% of all of its waste.

My suggestion is that if you live in one of these nonrecycling communities, go to a city council meeting and ask why recycling is not a priority. The city council can require the refuse companies to come up with a program. It's easy for households, and you can be the hero who started the ball rolling!

Home Energy

Insulation	Insulation will pay for itself in a few years.

Lighting	Change to LED bulbs and turn off lights (see Figure 48-6).
Solar power	Add solar panels to homes and businesses; the result: lower costs.
Appliances	Update to Energy Star efficient appliances.
Heating/cooling	Adjust thermostat at night or when away. Install a heat pump to improve efficiency.
Monitoring	Track your monthly power usage with The Energy Detective.

Figure 48-6: Comparison of Lighting Costs vs. Life Expectancy

Bulb Type	Least Efficient ————————————————— Most Efficient			
	Incandescent	**Halogen**	**CFL**	**LED**
	Energy used	Energy used	Energy used	Energy used
450 Lumens	**40w** $4.82/yr	**29w** $3.49/yr	**11w** $1.32/yr	**9w** $1.08/yr
800 Lumens	**60w** $7.23/yr	**43w** $5.18/yr	**13w** $1.57/yr	**12w** $1.44/yr
1100 Lumens	**75w** $9.03/yr	**53w** $6.38/yr	**20w** $2.41/yr	**17w** $2.05/yr
1600 Lumens	**100w** $12.05/yr	**72w** $8.67/yr	**23w** $2.77/yr	**20w** $2.41/yr
Rated Life	1 Year	1-3 Years	6-10 Years	15-20 Years
Lighting accounts for 20-30% of electric bill				

Estimated energy cost per year is based on 3 hours of use per day at 11 cents per kWh in an average single family home according to the Dept. of Energy

My dad actually tipped me off on this one, and it's another area where you can really make an impact. In the chart, you can see that a 100-watt incandescent bulb is equivalent to a 20-watt LED bulb! This simple change to LED light bulbs can make a remarkable impact both in energy use and your costs because lighting is typically around 15%-20% of your total energy bill!

Earlier I showed that for a 100-watt bulb, the difference between incandescent and LED could be almost $10/year. So imagine that if you replace ten light bulbs in your home, your savings could close to $100 a year, for 20 years = $2,000!

Insulation would be another easy target, even in newer homes, unless they are specifically designed for energy efficiency. And the nice thing is that you'll recover your investment probably faster than you think.

Solar power has a lot of opportunity and is better than in the past; there are many companies and groups that can give you a load of information if you ask.

Home and Garden

Lawn Care   Replace gasoline engine equipment with electrical products.

Lawn and garden is another opportunity area. I recently purchased an electric trimmer to replace my worn-out gasoline trimmer. It's *much* quieter, powerful, and gets the job done: just be sure to get a big enough battery based on how long it takes you to trim/edge your lawn. Greenworks has a full line of products—from leaf blowers to vacuums to lawn mowers and even snow blowers powered by electricity.

Home Care

Pests   Use environmentally friendly practices and pest control companies.
Use Chromated Copper Arsenate–treated wood for building.

Cleaning   Use EPA-approved "Safer Choice" cleaning and other products.
Wash clothes in cold or cooler water.

Social
Donation   Encourage sustainability in other countries; e.g., Heifer.

Education   Volunteer or promote educating children on what sustainability is.

What I really like about Heifer (and I'm sure there are others) is that it perpetuates sustainability through education and that it does so in places (including the United States) that really need economic development. Heifer's efforts contribute to that economic growth by using local natural resources and processes.

Education can also play a huge role in sustainability success for the future. The more kids understand the concept, the more they will gravitate to developing ideas that lead to sustainability actions. As I mentioned earlier, young adults entering the workplace today are already aware and are being active in sustainability planning in their companies and their communities; we just need to encourage and support those actions.

I am sure there are hundreds of additional easy things you can do to make an impact on sustainability. I'd like to hear from you; if you do so, we can post them on our website, Kickinggasandtakingcharge.com. And if you'd like to learn more about sustainability ideas, please check the ones listed in the upcoming Resources section of the book.

I'm not a tree hugger, and I'm not asking you to be. I'm asking you to do your part to keep the Earth as we know it for our kids and their kids. It's that simple.

Wrap Up

I don't profess to be an expert on sustainability after writing a chapter in this book. And while it is the end of this book, I don't want it to be "the end"; rather, I hope you'll let this book serve as inspiration for you to begin your own sustainability journey. And to that point, I made up this little reminder that "Sustainable is Maintainable if the Habits are Retainable." In other words, don't make your change a one-time-only change; commit to it. In my case, I've committed to recycling as many things as I can; now I take a small trash bag out every other week—everything else is recycled. I've committed to changing my light bulbs to LEDs and started with my outdoor flood lamps. I committed to an electric trimmer when I once used one powered by gasoline. I've committed to a hybrid vehicle. It's not hard—it just takes a little focus.

If you knew that the Earth would run out of oxygen in one year, wouldn't you commit to doing every recommendation offered up so that you could live beyond that one year? I think your grandkids and their kids would appreciate your promise to practice at least one or two ideas for that same reason. And hopefully what I have offered up is not too difficult of a commitment. If all of us do our part, we get movement, and improvement!

As I climb down off my soapbox now, I want to thank you for allowing me to share what I learned from the Tour. I hope it was entertaining. I hope it was educational. And I hope it was motivational. I know it was for me.

And thank you in advance for making an impact!

I'm still "Kicking Gas & Taking Charge" for my kids—forever; i.e. this is NOT… "THE END"

Resources-Sustainability

Here are a few places you can learn more about sustainability and gather other ideas on how you can "take charge" for the sustainability cause:

www.yousustain.com/footprint/actions (actions vs. savings calculator)

www.epa.gov/sustainability (everyone)

www.epa.gov/saferchoice/products (house-EPA listing of "Safer Choice" products)

www.conserve-energy-future.com/15-ideas-for-sustainable-living.php (households)

www.pinterest.com/explore/sustainability-kids (posters and ideas for kids)

www.shelburnefarms.org (for schools and education)

www.iamc.org/Resources/Top-Sustainability-Ideas (for businesses)

www.atkearney.com/sustainability/ideas-insights (for businesses)

www.sustainablebabysteps.com/going-green-at-home.html (around the house)

REFERENCE MATERIALS

Main Character Bios

Susan Jones (Arciero)

Role: Project Sponsor and Organizer; Electric Scooter Rider

Education: Furman University Graduate Masters Degrees in Psychology and Acupuncture

Published Author: *The Greatest Job on Earth: Extraordinary Parenting* and *The Hardest Job on Earth: Committed for Life*

Business: Owns Xenon Motor Company and Nashville Electric Scooter Tours
Owns Best Bangkok Tours in Thailand
Started Gardens R Us (growing gardens for consumers)

Resides: Bangkok, Thailand

Personality: Big Dreamer; wants to save the planet from human destruction
Purist for simplicity; loves new ideas
Accomplished acupuncturist for 18 years and alternative-medicine advocate
Mother of four daughters, all with singing backgrounds

Facebook: www.facebook.com/actuallysusan

Duane Leffel

Role: Nissan Electric Car Driver

Education: University of Cincinnati grad:
Metallurgical Engineering / Business
Accomplished speaker at Professional
Pricing Society Conferences

Business: Former Director for Nissan North American
Owner of Funexcursionsinabox.com travel guide

Resides: Nashville, Tennessee

Personality: 30-year career in Sales & Marketing for Nissan/
Caterpillar
Pragmatist and conservative planner, avid traveler,
motorcycle rider
Retired in order to join the Tour and to start a scooter
business with Susan
Father of two daughters (pictures on his Tour T-shirt)

Facebook: www.facebook.com/duane.leffel

Ben Hopkins

Nickname: "BenHop"

Role: A2B Electric Bicycle Rider

Education: Alderbrook High School, Solihull,
United Kingdom

Business: Professional Cartoonist
Consultant to A2B Electric Bikes

Resides: Maldives Island/Bangkok, Thailand

Personality: Accomplished bicyclist across Asia, Australia, Europe
Big Brit teddy bear with great sense of humor
Focused on daily tasks and very experienced on a bike
Guitar player and folk song singer; loves Bob Dylan

Facebook: www.facebook.com/benhophip

Ben Rich

Nickname: "Benswing"

Role: Zero Electric Motorcycle Rider

Education: Colgate Univ. Graduate in Physics
University of Maryland Graduate in
Secondary Science Education
Member Electric Auto Assoc. and American
Motorcyclist Assoc.

Business: Science Teacher and Sustainability Coordinator at
Montclair Kimberley Academy High School
Former Counselor at NASA United States Space Camp

Resides: Montclair, New Jersey

Personality: Huge motorcyclist and electric vehicle enthusiast
Very knowledgeable about electric vehicles
Recent 8,000-mile motorcycle trip around the United
States and into Mexico and Canada
Professional swing dancer dancing at locations around
the United States
Single, big smile, seeks press coverage on electric-vehicle

Facebook: www.facebook.com/benswing

George Wymenga

Role: Documentary Sound Technician

Education: N/A

Business: Sound man for commercials, films

Resides: Los Angeles, California

Personality: Single; experienced sound professional; quiet and
reserved
Does regular commercials involving Shaquille O'Neil

Facebook: None

Dominique Arciero

Role: Electric Scooter Rider, Singing Entertainer

Education: N/A

Business: Professional musician, singer, and songwriter

Former artist at Sony Records, with her sisters, known as The Lunabelles
Former member of sister singing group that toured with Kenny Rogers
Former President of Nashville Scooter Tours

Resides: Los Angeles

Personality: Daughter of Susan Jones (Arciero)
Free spirit who wrote and sang songs during the Tour
Smart, flirt, partier, capable, soft-spoken, world traveler
Along for the adventure and to further her career

Facebook: www.facebook.com/dominique917

Sean Scott

Role: Support Truck Driver and Logistics

Education: Biology student at Northeastern University
Studies at Bond University, Australia

Business: The Broad Institute
Former intern at Genzyme and Cubist Pharmaceuticals

Resides: Boston, Massachusetts

Personality: Native Hawaiian; avid hiker and traveler
Smart, curious, creative; a carefree and single hunk
Searching for an appropriate career and life path
Took Tour to bond with his father, to see the country, and for adventure

Facebook: www.facebook.com/kunu.scott

Rachel McCarthy

Role:	Children's "Being Green" Educator, Scooter Rider
Education:	Loreto College Coorparoo (Catholic School), Brisbane
	Bachelor's Degree in Creative Industries, Queensland U. of Technology
	Master's Degree in International Relations, Griffith University, Brisbane
	Master's Degree in Communications, Journalism, and PR, University of Queensland
Business:	UNICEF Education Officer, Phnom Penh, Cambodia
	Former Assistant for UNESCO, Bangkok, Thailand
	Former Course Coordinator and Lecturer at Thammasat University, Bangkok
Resides:	Phnom Penh, Cambodia
Personality:	Native Australian with international work experience
	Caring person who loves studying cultures and people
	Free spirit trying to find her niche in life; loved by all
	Never rode a scooter before the Tour
Facebook:	www.facebook.com/rachelmccarthy

Evan Scott

Role:	Documentary Videographer
Education:	N/A
Business:	N/A
Resides:	Los Angeles, California
Personality:	Young and creative; fun seeker, risk taker
Facebook:	www.facebook.com/evanscott

Stuart Scott

Role: Consumer Educator and Lobbyist

Education: Columbia University grad., Bachelor of Arts in Computer Science / Math University of Florida Master of Science in Computer and Information Science

Business: Founder of United Planet Faith and Science Initiative (UPFSI.org)
UPFSI aims to change attitudes and actions of humanity by saving the climate, biodiversity, and ecosystems of Earth for future generations.
Former consultant to Cambodian government on carbon emissions
Former U.S. Congressional lobbyist for global-warming legislation
Former IT and philosophy university instructor, software engineer, programmer

Resides: Honolulu, Hawaii

Personality: Intelligent and passionate about global warming and carbon-footprint issues
Regimented, outspoken, formal demeanor with "edge"
Wants to educate politicians and public about global warming via speeches/meetings
Involved his son Sean, who drove the support truck, as a bonding opportunity

Facebook: www.facebook.com/StuartGaia

Jonathan Becker

Role: Documentary Film Director

Education: Marketing Degree, Michigan State U.

Business: Director/Producer at BeckerFilm
Senior Producer/Director at Leo
Burnett-Greenhouse
Director/Producer for KickGas.TV and MADZILLA
Cofounder of The Jukebox Romeos

Resides: Chicago, Illinois

Personality: Creative, fun; laid back but focused on filming
Wants film to be successful, involved in Tour topics
Met Susan after doing creative work and training for
Nissan LEAF event
Goal: create breakthrough environmental documentary

Facebook: www.facebook.com/jonathanbecker1

Terry Hershner

Role: Electric Motorcycle Rider with Tour
(California)

Nickname: "Electric Terry"

Education: Engineering Degree at North
Carolina State University
Studied Environmentalism at U. of Central Florida

Business: Board of Directors at Electric Auto Association
Owner/Operator of Gas-Free Earth, a Nonprofit Org.

Resides: Santa Cruz, California

Personality: Well-known electric vehicle/motorcycle consultant
International long-distance motorcycle rally rider
Rides a Zero Electric motorcycle with his dog, "Charger"
Environmentalist who's lived "off the grid"

Facebook: www.facebook.com/terryhershner

ABOUT THE AUTHOR

Duane A. Leffel is a first-time published author who's no stranger to writing. He started writing in high school, where he wrote poetry and articles for the high school newsletter. His corporate career further honed his technical and editing skills, but it never diminished his creating writing skills; his annual Christmas letters poking fun at family events in the past year drew rave reviews from readers. And today, Duane uses these skills in developing Do-It-Yourself travel guides for people who want to set their own sightseeing agenda.

Duane is a former Nissan North America executive with over thirty years as a Sales and Marketing professional in the automotive and construction industries. As Director of Pricing Strategy for Nissan, Mr. Leffel was involved in numerous new vehicle launches and all Nissan and Infiniti vehicle pricing from 1999 to 2011. His career highlights include oversight of a $1 billion incentive budget, development of a $1 billion pricing-strategy proposal approved by top global executives, and the pricing of seven new-model vehicles during a corporate relocation from Los Angeles, California, to Nashville, Tennessee.

An accomplished speaker, Mr. Leffel has given speeches at the Professional Pricing Society and at pricing conventions on various topics. His ability to connect brand value, pricing and the four P's of marketing to a product everyone can relate to has made him a popular speaker at these events.

Duane lives in Nashville, Tennessee, as an eligible bachelor and is the proud father of two grown daughters. He has a degree in Metallurgical Engineering and a Business Certificate from the University of Cincinnati. Technically "retired," he recently launched a DIY travel guide website (fun-excursionsinabox.com) and enjoys riding his Honda Goldwing motorcycle when he isn't traveling or spending time with his daughters.